The world of our grandchildren

Bert Koene

The world of our grandchildren

Uitgeverij Aspekt

The world of our grandchildren

Amersfoortsestraat 27, 3769 AD Soesterberg, The Netherlands
info@uitgeverijaspekt.nl-http://www.uitgeverijaspekt.nl

Cover design: Mark Heuveling
Text layout: Pelle Vastenhoud

ISBN: 9789461537447
NUR: 400

For Erika, who asked so many questions that the answers became a book

Contents

Preface

Predictions are everywhere. Economists, financial experts, environmentalists, political commentators: they shower us with prognostications regarding the banking crisis, the climate crisis, the population crisis and the employment crisis. Their forecasts vary widely, but almost all agree that the future doesn't look bright.

Fortunately, experience shows that predictions made in times of real or imagined crisis tend to be too pessimistic. The extent of recent economic setbacks is more or less comparable to what happened during the oil crisis of 1973-1976. At that time, a drastic reduction of oil supplies from Arab states to Western countries that lent support to Israel in the Yom Kippur war, led to a huge increase of the price of oil. The Netherlands were especially hard hit. Since this country had delivered arms to Israel, it suffered a complete Arab oil boycott. The Dutch government saw a future where the national economy would wither away for many years and where the port of Rotterdam would forever lose its position as biggest oil port of the continent. The prime minister warned the Dutch people in a dramatic speech that the times of prosperity were over and gone. In reality, the economy bounced back within three years, Rotterdam experienced phenomenal growth and only elderly people still remember vaguely that there has been something like an oil crisis in the seventies.

Sticking to the example of Dutch history, it is not difficult to come up with events that were far worse than the oil crisis of 1973 or the prosperity correction suffered by the western world since 2008. I do not count times of war, they are in a category of their own, but we may for instance consider the early years of the nineteenth century. The Netherlands were at that time financially in a worse state than Greece is today. State income was just enough to pay interests due on loans. Any real expenditure increased the public debt. Industrial activity and trade were at a very low level, a large part of the population had no work and there was no prospect of better times ahead. It is no wonder that writings about the future of the country oozed gloom and doom. The country needed half a century to return to a state of acceptable economic and financial health.

Of course, such misery differs much from the rather minor drop in wealth caused by the banking crisis of 2008, but it is striking how present-day views of the future resemble Dutch treatises published in the dark times of two centuries ago. One again presents linear projections from the present into the future, with too much emphasis on economic indicators and quite blind to technological innovation and social change. My suspicion that such attitudes tend to produce unrealistically dim outlooks, made me decide to explore the future myself. This book is the result of that research.

A prevalence of positive sentiments can similarly lead to overly optimistic views. Twenty years ago, the end of the Cold War and a booming economy brought commentators into the mood for extremely upbeat predictions. Among other things, it was said that economic recessions were a thing of the past and that the western model of society based on the marriage of parliamentary democracy with capitalism constituted the terminal point of societal evolution.

Exploration of the future is interesting only if one tries to see at least several decades ahead. I decided to make it a century, and sometimes I will try to go even further than that. The title of the book speaks of the time of our grandchildren, but the reader may just as well think of great-grandchildren or even grandchildren of grandchildren. On such a time scale, matters of finance and economics are not of primary importance. They are only of interest if one tries to make a rather detailed forecast for just the years ahead. History shows that long-term ups and downs of the economy are only secondary phenomena, driven by social developments and progress in science and technology. On really long time scales it is in particular the second factor, progress in tech-science, that counts.

Science and technology are root causes of change and progress, but for that very reason they are also an important source of problems. The three most important questions of our time may in fact be seen as consequences of technological progress. First, the explosive growth of the world population is the result of better health care and increased food production. Second, the pace of economic globalization has enormously increased by the availability of world-wide tele-communication, cheap and fast transportation and the digitization of information. Third, the climate crisis is the result of the burning of huge quantities of fossil fuels in power stations and means of transport. These three issues will dominate world politics in the 21st century. I will devote chapters to each of them.

The influence of science and technology on the daily life of individual citizens and on society as a whole will in the coming centuries probably keep growing at ever increasing speed. Therefore, I will in one chapter try to look about five hundred years ahead in some areas of science and technology, but I must admit that prognostication becomes a very uncertain affair when one ventures that far into the future. In all other parts of the book I will concentrate on the 21st century.

Quite a number of authors have tried to explore certain aspects of life in the 21st century, such as climate change or technical innovations or politics, but broad ranging investigations of the future are rare. For me, the central issue was: what can be said with reasonable certainty about the living conditions of the average citizen in the next hundred years. I will discuss everything that touches on that question.

It is my aim to present an objective inquiry into possible future developments, without utopian or apocalyptic seasoning. Still, I am aware that my professional background will influence the way in which I present certain facts and trends. I have worked for thirty years as a research physicist in several countries. Ten years ago I switched to something entirely different. I became a historian.

Only in the last chapter will I leave the neutral position of reporter of likely or possible developments.

I will then place three danger signs, but it would be an exaggeration to consider my warnings as hints for survival. That would suggest a much darker outlook than I paint it. In the words of a weather forecast, my findings boil down to 'increasingly clearer skies with some scattered clouds and a risk of an occasional thunderstorm'.

Sense and nonsense of predictions

The future is not predetermined

Classical prophecy assumes the future already to be fixed. Fortune-tellers and soothsayers claim to 'see' aspects of that predetermined future in tarot cards, coffee grounds, the palm of a hand, intestines of birds, et cetera. One hundred years ago the belief in predestination was not yet in explicit conflict with scientific knowledge. Natural scientists of the 19th century saw the universe as an extremely precise clockwork. From the moment it was put in motion, the laws of nature determined exactly what would happen at any later moment. Once a gullible person was ready to accept that a soothsayer saw his future in a crystal ball, there could not be any doubt about the correctness of the vision. Science itself said the future was already entirely fixed, so if somebody was truly able to look into the future, he or she was certain to see the 'right' future.

Today we know that the future is *not* fixed. In the first decades of the 20th century physicists discovered that the world is not deterministic on the scale of atoms and molecules. The number of different ways in which this micro-world can evolve from moment to moment is gigantic. From our point of view, the differences are extremely tiny, but they may add up step by step, so that in the end two possible sequences of micro-events can lead to notably different versions of the future. The atomic world obeys the laws of quantum physics. These laws only know of possibilities, not of predetermined certainties. The choice between possibilities is in the quantum world only made at the very last moment, that is, when the future becomes *now*. It is a selection process that is governed by probabilities and blind chance. In the words, if one believes in God, one must accept that He plays dice.

To give an example, suppose two atoms collide in such a way that there are three possible outcomes: both atoms disintegrate, one of them disintegrates or both stay intact. In such a case the rules of quantum physics determine how probable each outcome is. If, for instance, the probabilities are 5%, 15% and 80%, then we know for sure that on average both atoms will survive such a collision in 8 out of 10 cases, but for any individual collision it remains a matter of chance whether it will belong to that 80%, or to the rarer cases where one or both atoms disintegrate.

This fundamental randomness remains hidden to us in our daily lives, because we experience the world on a macroscopic scale. Everything we can detect with our senses, every object, every phenomenon, involves billions upon billions of atomic particles. What we experience, is the *average* behaviour of all those particles, which is determined, strictly and precisely, by the probability rules of quantum physics. We are not aware that the behaviour of *individual* atoms and molecules is governed by blind chance. Still, quantum uncertainty influences the course of human history, one reason being that the molecular biology of our body touches the realms of quantum chemistry and quantum physics. For example, signal transfer between our brain cells depends on the transport of calcium ions in the synapses, the contact points between nerve cells. This ion transport is sensitive to quantum effects. With the rise of nanotechnology the average citizen will in the near future even become a routine user of applied quantum physics, although only few will be aware of that.

Sources of uncertainty

I brought up the subtle quantum haziness of the world at the scale of its building blocks, atoms and molecules, in order to show that the basic assumption of traditional fortune-telling – the existence of a predetermined future – is fundamentally incorrect. However, quantum uncertainty is not the main reason why the future of mankind is so difficult to predict. There are more mundane obstacles that count more heavily. I enumerate them in order of increasing importance as a source of uncertainty.

First: incomplete knowledge of the present state of our society, although that is where the whole exercise of making predictions starts.

Second: insufficient understanding of the mechanisms that determine social, cultural and economic developments within a society, while such understanding is essential if one wants to argue from the present to the future.

Third: pure chance in the world around us. A futile incident can lead to an avalanche of events with enormous consequences.

Fourth: the whims of human nature. Impulses and emotions influence everybody's actions.

I give an example of the role of mere chance. Two executives of a big company get stuck in the elevator of their hotel, which makes them miss an important board meeting. With their votes investment plan A would have been accepted, but one now goes for plan B, which soon brings the enterprise into bankruptcy. A domino effect among suppliers then destabilizes an entire branch of industry. This harms business and consumer confidence to such a degree that the economy of the country falls into recession. All this because of a jammed elevator door.

Whimsical behaviour of people is by far the most important reason why the course of history cannot be predicted in any detail. I give an example from the year 1789, the first year of the French revolution. It was purely by bad luck that on a Parisian street a quarrel with far reaching consequences erupted between two hotheads. These gentlemen would not have run into each other if one of them had not left home at an hour that for him was unusual.

Of course, the fact that absolutely certain, detailed predictions are not possible, does not mean that it would be fruitless to look for the most likely general course of events. Nobody could predict that the French Revolution would start with the assault on the Bastille on the 14th of July 1789, but any sensible observer must have seen that the widespread discontent among the population was building up towards an inevitable eruption. This is how it often goes with forecasts: it appears very likely that a certain type of event will happen, so it makes sense to analyze its possible consequences, but the time of occurrence of the event, the form it will take and the precise chain of secondary events it will cause, cannot be foreseen.

The behaviour of individuals or small groups of people is highly unpredictable, but a collective trend, a general direction of change in attitudes within a large group of people, is in principle easier to keep track of. A novelty strongly rejected by some people in such a group may be greeted enthusiastically by others, so extremes average out. What remains is a trend, a general attitude that does not depend on personal idiosyncrasies and that can, therefore, to some degree be foreseen. It is for this reason that developments in a democracy are more predictable than in an autocratic state, where everything depends on the moods of one man or woman.

Unreliably trends

The most striking feature of predictions made in the past is that they were so often so completely wrong. The main reason for that poor record is that the authors allowed themselves to be carried away by prevailing moods of their days. When people in the West trembled for the Red Danger, there were many who, like George Orwell with "Animal Farm" (1945) and "Nineteen Eighty Four"(1949), predicted a pitch dark future of world dictatorship. In the fifties of the last century others were infected by a mood of exuberant optimism about the material blessings that 'splitting the atom' and other feats of science and technology would bring humanity.

A solid background in science is no guarantee at all that somebody will correctly anticipate triumphs of science and technology, even if these are already in the making. Lord Kelvin, a famous physicist and chairman of the Royal Society of London, the most distinguished scholarly society of the 19*th* century, knew for sure that 'radio has no future', 'X-rays will prove to be a hoax', and 'no aero plane will ever be practically successful'. The last statement is from 1902, just one year before Orville Wright went a few meters up into the air with the flying machine he had built with his brother Wilbur. Lord Kelvin's scepticism could be attributed to his age – he was 78 years old when he gave his opinion about aviation – but even the young inventors themselves were blind to the future impact of their work. Five years after his memorable flight, Orville declared: 'No flying machine will ever fly from New York to Paris'.

Remarkably good are the predictions made by an otherwise insignificant American engineer named John Elfreth Watkins. In 1900 he contributed an article to the magazine *Ladies' home Journal* under the heading 'What May Happen in the Next Hundred Years'. About half of his 28 predictions have turned out to be essentially correct. He anticipated television and even described its principle of operation quite well: 'Man will see around the world. Persons and things of all kind will be brought within focus of cameras connected electrically with screens at opposite ends of circuits, thousands of miles at a span.' A miss was: 'There will be no C, X or Q in our everyday alphabet. They will be abandoned because unnecessary'.

Even more interesting is a recently discovered list by the British scholar sir Robert Boyle, often seen as the first modern chemist and in any case a pioneer of experimental science. He must have composed the list around the year 1660. It is not so much a prediction, more an enumeration of inventions that seemed very desirable to him, such as 'the art of flying', 'a ship to sail with all winds' and a 'perpetual light'. If we are a little bit flexible in the interpretation of his phrasings – for instance, considering electric lighting to fulfil Boyle's wish for 'perpetual light' – we might say that 22 of the 24 items on his wish list have by now been realized. That is an extremely high score. Is it pure chance, or did Boyle possess a sharp understanding of what was in principle possible within the laws of nature and did he put on his wish list only dreams that he felt to belong to this class?

Turning our attention to writings expressly meant to be prophetic, we note that the extremes of apocalypse and utopia have been most popular. That is not surprising. Everybody is sensitive to the predominant sentiments of his or her day, and it is easy to consider them as immutable, with the result that they become overly emphasized when one tries to look into the future. If someone decides to buy stock after prices have been rising for quite some time, then he is not just listening to his instinct to follow the herd, he is also driven by belief in the durability of trends. The longer the prices of stocks or any kind of commodities have been rising, the deeper the conviction that they can only continue to rise. Housing markets in several countries have suffered repeatedly from a similar misconception.

In general it is of course a sensible start to assume that tomorrow things will be about the same as today. It is the best possible prediction about any topic if we are not able or willing to investigate it. Somebody locked up in a cave without any knowledge about the weather above, may do quite well as a weather forecaster by simply announcing that tomorrow's weather will be very much the same as today. However, this strategy does not work if one wants to look further ahead. One can then no longer trust that present circumstances and present trends will continue unchanged. Simple extrapolation is no longer good enough, time graphs with only straight lines do not suffice.

Identification and extrapolation of trends is not an exact science. Wishful thinking plays a role, as is often the case when one has to interpret incomplete data. Long-term trends may be masked by short-term movements back and forth in politics, fashion, arts, et cetera. Moreover, an apparently steady long-term trend of rise or fall may in fact be only a mounting or falling segment of a very slow oscillation between ups and downs. There are also many examples of sudden explosive growth, which must either burst like a soap bubble – like the Dutch tulip madness in the 17th century, when a rare tulip bulb could have the price of a house, or the internet bubble on the stock market between 1997 and 2000 – or be stopped in their growth by the finite amount of available resources or by physical limits.

Being stopped by a limit of the last kind will be the fate of Moore's Law, which says that the density of circuit elements on computer chips, and thus computing power, doubles every one and a half or two years. This statement, which is not really a 'law' of nature but a loose prediction, has for a remarkably long time turned out to be true, but is bound to get into trouble soon. The reason is that the miniaturization of the lithographic patterns on microchips is already approaching the scale of atoms and molecules, where quantum effects make classical electronic circuitry useless. Novel kinds of electronic components and circuits will have to be invented, which will for some time slow down progress to below the pace predicted by Moore's Law.

Moore's Law describes exponential growth. A quantity is said to grow exponentially if it exhibits constant *relative* increase: growth by always the same *fraction* per unit of time. It is the kind of growth enjoyed by capital that earns a fixed yearly interest, provided that each year's interest is added to the capital. It is the mechanism of interest on interest that leads to exponential increase. The power of this kind of growth is demonstrated by the fact that a smart mobile phone of 2014 packs more computing power than NASA had at its disposal for putting two astronauts on the moon.

The biggest obstacle to making reliable predictions is the possibility that a stable trend will suddenly be broken by an unpredictable event, something in the class of a meteorite suddenly hitting the earth. Tame economic trends and even cases of explosive growth – always doomed to die – can in principle be handled, but obviously nothing can be done against what is unforeseeable. Of course, it must not always be bad news when a forecast turns out to be wrong (except maybe for the forecaster who wants nothing else than being right.) For example, early in the 20th century the prediction of exponential growth of the world population appeared very realistic, but present data reduce it to a doomsday fantasy, as we will see in a later chapter.

Initiators, followers and shapers of public opinion

Many seemingly important problems of today are of little lasting consequence. That includes the banking crisis of 2008 and the euro crisis which followed it. Certainly, the hundreds of billions of public money that had to be spent on failing banks were a bitter pill for the tax payer. The cost of supporting ailing banks and improvising measures to keep the euro project alive – needed because the construction of safeguards had been neglected at the time the common currency was launched – has thrown many Europeans back to the prosperity level of five or ten years earlier, but that is a triviality in the course of history. The fact that for some years we had to endure the financial pain of corrective measurements because governments and financial institutions did not have their affairs in proper order, will have little effect on how the world of our grandchildren will be ticking fifty or one hundred years from now, because the entire episode did not result in fundamental changes to the financial and economic system.

Quite generally, events in the areas of finance and economics are seldom of long-term significance. The world of industry, commerce and money is essentially a follower of developments, not a driving force. Changes – mostly for the better – in the conditions of human life have in the past centuries mainly been due to new technological inventions, and these were the result of new scientific insights. Engineering and science have become the engines of human history. With some delay, industry and commerce profitably turn the inventions into new products. The fact that today many of the largest companies of the world are active in the fields of informatics, computing and telecommunications, shows that technological innovation can within a few decades lead to enormous shifts in the world economy.

Science and technology are at the root of change, but their practitioners do not themselves wield much influence. Politics, business, industry, religious organizations and the media play a larger role in the shaping of public opinion, also on the subject of acceptance or rejection of new technologies. I will repeatedly examine the attempts from various quarters to steer, promote or brake scientific and technological progress.

Just like today, hunger for power, self-interest of organizations, ideological preoccupation and other irrational motives will in the next one hundred years again define the positions taken by the various parties on the playing field of an open society. It is very hard to predict anything about trends in such power plays. In other words, it is easier to identify technological inventions with great potential than to predict how society will react to them. Still, it is that reaction which determines if and how new technology will actually be implemented.

So far I have tacitly assumed that European countries will in the next one hundred years still enjoy some sort of democratic government. For a country under autocratic government it is much more difficult – maybe nearly impossible – to predict what the attitude will be towards future scientific and technological progress. In such a country there is no competition between ideas and insights, no possibility for the pre-judices of one party to be corrected by the other. All depends on the whims of the ruler or his small clique..

Point of departure: things could be worse

We are doing better than ever before. Fifty years ago the world lived under the threat of imminent nuclear war (the Cuba crisis 1962). Today, there are several wars going on, but those are regional conflicts or civil wars. They are items of the evening news, not a reality at the doorsteps of the citizen of a western country. Even along the line of friction with Russia, where European peace could conceivably be disturbed, it seems unlikely that violence on the scale of warfare will erupt in the coming decades, thanks to the fact that both sides are aware of their economic interdependence. Since 1945, so already for seventy years, there has been no war in Europe, except for the Balkans. That is unique in history. For comparison: when one looks back from 1945 to the 14th century (the earliest time with more or less reliable information for large parts of Europe), one encounters in those six centuries on average two new wars in every year.[1]

It is extremely unlikely that the peace between and inside the nations of the European Union will be broken in the coming century. The Union has intertwined the fortunes of its member states to a high degree, starting with their economies and then proceeding to many other areas. The result is a structure of stabilizing connections, which has been erected al-

1 Numbers mentioned by Steven Pinker in his Tinbergen lecture in 2013. (www.science.leidenuniv.nl/index.php/tinbergen/lezing-2013)

most without the citizens of the member states being aware of it, and without much coherent planning by their political leaders. From incident to incident, the European politicians have again and again been forced into approximately the right direction by the power of circumstances. We should be happy with that result, because it guarantees peace and a certain degree of harmony within Europe, but the way in which it was reached, is not a source of pride for many politicians.

What the fate of the euro will be, whether some countries will leave the euro zone or not, in how far the banking system will be rearranged, those are all questions of little importance when we want to look one hundred years into the future. I already stated that financial and economic matters are in themselves of little importance for the long-term course of history. They only become of interest if they are the cause of really decisive events, like wars. Since it appears almost excluded that the euro crisis will lead to outright hostility between member states, the details of how the euro zone is going to cure its problems will have little influence of the state of Europe and the world some decades ahead.

The prospect of peace inside the European Union does not mean that it will stay quiet in the neighbourhood. Russia is far from being a stable country and, apart from the Baltic States, Georgia and Armenia, all other parts of the former Soviet Union are artificial entities. Ukraine and Belarus, the most western of these quasi-states, may remain sources of unrest for many decades to come. What has been happening in Ukraine since 2013 – heating of the fire by Russia, but not to the point where it would provoke serious intervention by the West – will probably keep repeating itself many times, also in Belarus. In the end, these two entities will either become parts of Russia, like in czarist times, or find protection under the umbrella of the European Union.

War between European countries and Russia appears unlikely, but very unpleasant episodes are to be expected. The preference of the Russian electorate for a so-called strong man – at this very moment in the person of a primitive nationalist with a dictatorial agenda – in any case does not bode well. Nevertheless, one can hardly overestimate the value of the prospect of lasting peace *inside* the borders of the European Union, a prospect that has never before in history been so realistic.

Something remarkable has occurred world-wide as well. A single day without war anywhere on earth is still a utopia, but there has been no direct fighting between major world powers since sixty years,. That is just as novel as the long peace in Europe. There is in fact more good news of a global scale, although it must be said that those positive developments are accompanied by some less pleasant drawbacks. It is the well-known phenomenon that success often comes with harmful side-effects.

The first success story concerns the agricultural revolution that took place between 1960 and 1980. The introduction of new varieties of crops, fertilizers, pesticides and new irrigation techniques made it possible to double or even triple yields per acre. The result is that countries like China, India and Indonesia can now feed their own people. Although problems in the *distribution* of food remain, the danger of a structural shortage in the global food *supply* no longer exists, even if the world population would still increase substantially. On the other hand, the often drastic methods of maximizing crop yields, like excessive use of fertilizer and pesticides, do serious harm to the environment, as does the extensive deforestation in several countries to increase acreage. Still, the balance sheet of the agricultural revolution shows a large positive return.

The increase of life expectancy through better nourishment and better medical care is the second great achievement of our time. In countries where favourable conditions have been in place for some time already, further gains are of course being made only at a modest rate, but in many parts of the world strongly improved living conditions are a thing of the last decades. In those regions, the life expectancy of a newly born has been rising fast, especially because the death rate in the first year of life has been strongly reduced.

This is of course a happy development, but the consequence is that the world population has been growing at an alarming rate. Continuation of that growth is the most elementary disaster scenario one can imagine, given that almost all kinds of resources are finite. Will the explosive growth of the world population continue until global famine and wars for water, land, raw materials and energy will put a miserable end to it, or are we just witnessing a temporary growth spurt, which will pass into a more moderate evolution of population numbers? This is the first question that anybody exploring the future must face. I will fulfil that duty in the next chapter.

The third of the three big, intrinsically positive developments is the rapid growth of world-wide interaction and interdependence in many areas of activity, as a result of the fact that communication and transport technologies have eliminated geographical distance as a significant obstacle. It is a broad phenomenon, but economic globalization is the most important instance. It causes a fast rise of living standards in many countries that until recently were classified as underdeveloped countries, mainly in East Asia but also in South America and a few African States.

Also in this case it is true that too big a success may be the shortest path to catastrophe. Worse even, population growth and globalization tend to amplify each other's negative effects. When not only the world population is growing fast, but at the same time the average citizen can afford ever higher levels of consumption, resources will be depleted even sooner and harm to the environment will be even more severe. The adverse effects of the agricultural revolution are small compared to what global growth of industrial activity, trade and consumption of goods could do to the environment.

Will mankind spoil its own future by the damage its pursuit of prosperity causes? I will indeed have to devote a rather depressing chapter to climate change, but that will be preceded by two mostly optimistic chapters about the unique changes occurring in our time regarding the growth of the world population and the increasing interdependence of the welfare of nations. The word unique really applies, because there is a change taking place from old structures and relations to new ones that offer better chances to more people.

Scientific and technological developments are the root causes of population growth, globalization and climate change, but how controlled or uncontrolled and how fast these processes occur, is determined not by science but by cultural, political and economic forces. Therefore, science and technology will remain in the background in the first part of this book, where I investigate the 'big three' of present world problems. They will take centre-stage in the second part, where I will concentrate on the theme of tech-science as the driving force of human history of the coming centuries.

Growth of the world population

The end of growth is near

In 1972 the report of the Club of Rome appeared.[2] The Club of Rome was an informal gathering of private citizens from all over the world who feared that mankind was destroying its own future. The report essentially was a signal of alarm, spelling out that population, food production, industrialization, waste production and use of non-renewable resources were all rising almost exponentially. Exponential growth happens when a quantity keeps growing by a fixed percentage. The title of the report, *The limits to growth*, pointed to the fact that such a process must sooner or later hit a ceiling.

In the sixties of the 20th century the global population had been growing by 2.1% annually. The Club of Rome believed it to be almost excluded that the growth rate would decrease before the year 2000, even under the most optimistic suppositions about future fertility rates. An annual increase of 2.1% would double the world population in 33 years. In 1970 approximately 3.7 billion people inhabited the earth. An unchanged exponential growth rate would make that 7.4 billion by the year 2003.

It has meanwhile become clear that the Club of Rome was too pessimistic. In 2003 the earth was pop-

2 D.L. Meadows, *The limits to growth. A report for the Club of Rome Project on the predicament of mankind* (Universe Books, New York, 1972).

ulated by 'only' 6.4 billion humans, a full billion less than feared.[3] The number of 7.4 billion will according to the last United Nations estimate be reached around the year 2016. Since publication of the report of the Club of Rome, the world population has been growing almost linearly. That means that the rate of growth is about constant in *absolute* numbers, instead of the constant *relative* growth feared by the Club of Rome. In terms of relative growth, one sees a trend of strong decrease: the rate of 2.2% in 1963 (the highest annual growth rate ever) decreased in the time of half a century to 1.1% annually in 2012.

In 2006 the United Nations announced for the first time that population growth was weakening also in absolute numbers. This trend appears to be continuing. In case of a modest further reduction of fertility rates in countries where big families are still common, recent studies (2012) predict that population growth will almost come to a halt in the second half of the 21st century, at a world population of 10 to 11 billion souls.

Why population growth stops

If demographers are right about their assumptions on future birth rates, the end of the population explosion is in sight. But are they? Reassurance can be found by studying the numbers of individual countries or regions, instead of only looking at the total world pop-

3 All population numbers in this chapter have been taken from the United Nations data base 'World Population Prospects: the 2012 Revision'. It is freely accessible at http://esa.un.org/unpd/wpp/.

ulation. In our time almost all parts of the world pass through great demographic changes, but these changes did not start everywhere at the same time and don't everywhere proceed at the same pace. The interesting fact is that some countries and regions have already passed all stages of their demographic transition, and that the outcome is in all cases the same: the population has become much bigger than it used to be, but stays almost stable at that high new level or even starts to shrink slowly. This observation justifies the hope that also in regions where the population is still growing strongly, stabilization will finally occur.

That hope becomes confidence if we don't just examine graphs and numbers, but also look for an explanation of the fact that different regions of the earth are in our time – second half of the 20*th* century and first half of the 21*st* century – going through almost identical demographic changes. Inspired by an investigation of rates of birth and death in industrialized countries in the 18*th* and 19*th* century, the American demographer Warren Thompson has already in 1929 constructed a model that explains the present demographic shifts quite well.[4]

Thompson distinguished four phases in demographic development. The first phase represents the pre-industrial society. High death rates due to poor nutrition and hygiene, from time to time aggravated by famine, epidemics and war fare, were approximately balanced by high birth rates, resulting in little or no growth of the population. The high birth rate had socio-economic reasons. Children were cheap labour

4 W.S. Thompson, 'Population', *American Journal of Sociology* 34 (1929), p. 959–975.

already at a very young age at the farm or the small family enterprise, or they earned a few pennies in a factory as long as child labour was allowed. And when they were old, parents would need children around to take care of them. These benefits outweighed the cost of feeding and clothing the children.

In phase two the death rate falls thanks to better nourishment, hygiene and health care. Europe entered this phase already in the course of the 19th century, first in the northwest of the continent, later in the south and east. Because birth rates stayed high, the European population began to increase strongly. Nowadays there is scarcely a spot on earth were the death rate has not been significantly reduced.

In underdeveloped countries, the reduction of death rates is almost synonymous with reduction of infant mortality. The world average of the number of children dying in their first year has fallen from 126 per thousand in 1960 to 45 in 2010. Averages still hide big differences between countries. In Japan and Sweden only two out of one thousand infants die, but in some African countries infant mortality is still around ten percent.

The mortality rate decreases 'all by itself' as soon as food supply and health care reach proper levels. People don't have to do anything else than make use of the improved facilities. Pushing back the birth rate is much more difficult. It requires considered choices by a population with some degree of education, living under a government capable of maintaining a reasonably ordered society. Countries with lowered death rates as a result of progress in agriculture and medical care, but with a stagnant social and political system, which prevents birth rates from dropping, have to face very rapid population growth. This is what happens in parts of Africa today. In 2012-2013 that continent was accountable for half of the growth of the world population. Nigeria, the most populous African state, today has 70 million inhabitants. If the present growth rate is maintained, there will be 730 million Nigerians by the year 2100. That is the same as the number of Chinese back in 1965.

In phase three of Thompson's model the pace of population growth slackens because the birth rate starts to decrease. In almost every part of the earth except Africa mankind is already far into this phase, or has even reached the fourth and final phase where rates of birth and death are approximately equal, so the population does not grow anymore.

What is the incentive for women to bear fewer children? Access to contraceptives helps of course, but an upgraded status of women and changes in the economic value of children for their parents are more important. If child labour is banned, if children have to attend school for many years and if they move from the village of their parents to a city to find work, then it is no longer profitable to have children. Instead, they become a financial burden. A couple wishing to be part of consumer society can in such circumstances not afford to have lots of children. The changes necessary for the reduction of birth rate can be summarized in a few key words: availability of education, urbanization, establishment of a socio-economic middle class.

Fast change

Birth rates can fall with stunning speed. The fertility rate, that is, the average number of children per woman, dropped in Bangladesh from 6.6 in 1975-1980 to 2.3 in 2005-2010. In Iran it fell over the same period from 6.3 to 1.9. The world average has since the report of the Club of Rome (1972) fallen from 4.5 children per woman to 2.5 in 2012. That is a decrease of 40%. Among the countries where the fertility rate decreased at least by that percentage, one finds countries that many people still associate with enormous swarms of children, like India, Mexico, Indonesia and Egypt.

The end to population growth has already been reached in Europe, East Asia (China, Japan, Korea), Iran, Argentine. It remains to be seen in how far the maintenance of an equilibrium between births and deaths is a self-regulating process. Governments will anyhow have tools and means to steer birth rates, softly with instruments like child support money or rigorously with measures like the one child policy of China.

At present there seems to be a tendency towards a slight population decrease in countries which have reached Thompson's final phase. If no women would die before reaching the reproductive age, the fertility rate would have to be exactly 2 in order to maintain a constant population size. Accounting for actual death numbers, one finds that the world average of the rate of fertility should at present be 2.33 to keep the world population at a stable level. The actual number of 2.5 in 2012 is encouragingly close to it.

In highly developed countries, where infant mortality is low, the replacement value is not 2.33 but 2.1. Actual fertility rates in Japan and Europe are far smaller. The average in Europe was in 2009 only 1.6. If that remains unchanged, the European population will rapidly decrease, unless a large number of immigrants is admitted. The most recent estimate of the United Nations for Europe including Russia is a decrease of population from 740 million in 2010 to 640 million in 2100.

China is a special case. The one child policy has brought the numbers of older versus younger people out of balance (and has led to a shortage of girls). The country will see a strong, destabilizing shrinkage of the population. Worse, relatively few economically active people will have to support large numbers of elderly citizens. In 2020 the mean age in China will be higher than in the USA and around 2030 China will even outdo Europe in that respect. As yet nobody knows how to solve this. It is in any case certain that the Chinese economic wonder will suffer severe setbacks as a result of the abnormal age structure of the country.

Since the demographic transition model of Thompson has been found to be valid in many different parts of the world, it is to be expected that the African nations presently staying behind will in the end also pass through all phases of the model. The birth rates in those nations will finally drop, causing the African population explosion to peter out. In short, it is my conclusion that the doom scenario of a world population that keeps growing until it hits the limits of resources, will not become reality. Late in the 21st century the earth population will stabilize at the level of approximately 10 billion people and may later even start to decrease slowly.

We are in fact witness to an extraordinary event in human history. We live in the last phase of a demographic transition that started two centuries ago in the northwest of Europe.

Before that time, the number of people on earth grew or fell by events outside of human control, like the availability of much or little food as a result of the weather, or the appearance and disappearance of epidemics. Almost hidden under the much more pronounced short-term fluctuations between good and bad times, one can in this early period already detect a slow long-term increase of the population as a result of the fact that man learned to manage his environment better, but the big growth spurt of the world population – from nearly one billion in 1810 to seven billion in 2010 – was triggered by better health care, implementation of a sanitary infrastructure and stable food supply.

It is important to note that in many so-called third world countries the decline of birth rates has started quite soon after death rates fell. Often the time lag was just a few decades. In Europe it had been a quite different story. It is where the demographic transition started first, but in Europe it took nearly two centuries to complete the whole process. Birth rates in East Asia and South America could follow the decline of death rates much more quickly, because these regions were partner to the process of globalization, at first mostly in economic terms, but then also culturally.

Life in urban regions in those countries is strongly influenced by economic, social and cultural interactions with highly developed Western countries. One could say that they imported the socio-economic conditions for lower birth rates from the West.

If the fertility rates in the big, populous countries of the third world would not have started to decline so quickly, the population explosion would not be ending at about 10 billion but at a much higher number. So we must thank globalization – the topic of the next chapter – for the fact that the earth population will not grow to an unmanageable number.

Globalization

Continuation of an old phenomenon

Now that we have come to the reassuring conclusion that the growth of the world population will come to an end in the course of the 21st century, we can turn to the question *how* the average world citizen will live in the next hundred years. Globalization will take to the centre of the stage.

I will not discuss in this chapter the unpleasant side effects of the stormy growth of the world economy through globalization. That I will do in the chapter on climate change. The reason for this choice is that those ill effects are *not* intrinsically connected to the increasing interdependence and integration of the world economy. The almost unlimited emissions of greenhouse gases in the past decades are not part and parcel of economic globalization, but the result of a socio-political failure, viz. the unwillingness of voters and politicians to take the right decisions.

For many people globalization equals economic globalization, the trend towards a world-wide web of connections, partnerships, dependences, treaties and agreements in business, trade and finance. It is, however, a much broader phenomenon of mutual influence, sharing and alliances between regions and societies. It is a process of economic, political and cultural integration which has been going on for a long time already, but which has lately been strongly accelerated by the advent of technologies that have eliminated geographical distance as an obstacle to close interaction.

First of all, transport of people and goods over long distances has become cheap and fast. These days it costs less to ship a container from Shanghai to Rotterdam than to transport it further from Rotterdam to Stuttgart. Second, telecommunication has been developed to the point where instant interaction by voice, image and writing is possible between almost any two cities on earth. Third, digitization of information makes it possible to access business data from any place. More tools that make distance irrelevant will become available, for instance, improved implementations of the idea of three-dimensional printing will make it possible to send copies of material objects to any place.

What we nowadays call globalization is a natural continuation of earlier steps towards integration. When mankind started agriculture about 10,000 years ago, life in nomadic clans was replaced by the more elaborate social structures of villages and small towns. Next, local chieftains and war lords subjected entire regions and went on to establish hereditary rule over their lands. These kingdoms, duchies, sultanates and other dynastic territories in due course were transformed into the nation states of today, often after further mergers. These early moves towards integration differ in one important way from the process of globalization of our time. While in earlier centuries the clash of arms was almost always involved, the modern drive towards integration and interdependence is mostly a peaceful process.

Earlier episodes of economical globalization certainly were not of high ethical standards. In the late 19th and early 20th century, the heydays of colonialism, economies and cultures became connected on a truly global scale, but in the asymmetric manner of masters and servants. It was a time when the wealth gap between Western Europe and its former North American colonies on one side and the rest of the world on the other side steadily widened. A bit further back in history, the slave trade must also be counted as part of the globalization process.

Boisterous growth of world trade is not a new phenomenon, there have been earlier periods of rapid increase. Between 1870 and 1913 the world economy was growing at a pace of 2 or 2.5% annually, which was three times faster than in the fifty years before. At the start of the First World War, the international trade of goods accounted for 12% of the gross national product (GNP) of the industrialized countries. After a prolonged slump as a result of the two World Wars and the Cold war with Russia, that high level was only reached again in the seventies of the 20th century.

Typical of the surge of globalization since the two last decades of the 20th century is the element of work sharing over different locations on the globe, often in such a way that companies keep their main office and departments of research and development in a highly developed country and set up production facilities in countries with low wages. In earlier times such a globe-spanning business model was not possible because instant communication and means of cheap and fast transport did not exist.

Globalization is the main cause of the erosion of national borders. The free flow of capital, goods and services requires the harmonization of rules and regulations in many areas, and intensive contacts between people in different countries automatically result in socio-cultural influences flowing back and forth, from frivolous subjects like cooking recipes to grave matters like women's rights.

Interdependent economies

Economic globalization makes the national economies of countries over the whole world more and more dependent on each other. Nowadays the most important trend is for highly developed economies to become interwoven with the economies of less developed countries through work sharing, direct investments and the dismantling of trade barriers. On the whole, this appears to be profitable for both sides. I give some numbers.[5] The annual growth rate of the GNP of poor countries which have participated in globalization since around 1980, has accelerated as follows: 1.4% in the sixties of the last century ; 2.9% in the seventies; 3.5% in the eighties; 5.0% in the nineties. The performances of the big, fast growers China and India have contributed heavily to these rather nice average numbers, but still, it remains true that nearly all underdeveloped countries which took

5 The quantitative data in this paragraph were taken from D. Dollar and A. Kraay, 'Trade, growth and poverty', *Finance and Development* (a quarterly journal of the IMF), nr. 38 (September 2001), accessible at www.imf.org/external/pubs/ft/fandd/2001/09/dollar.htm.

part in the process of globalization realized significant growth rates.

In the same period the economic growth rate of the rich countries fell from 4.7% in the sixties to 2.2% in the nineties. Poor countries which played no part in economic globalization fared quite badly: the annual growth rate decreased from 3.3% in the seventies to just 1.4% in the nineties.

The fact that a number of countries in Asia and South America are in the process of narrowing the gap with Western industrial countries does not mean that the world is moving towards an entirely novel distribution of wealth. In the case of Asia, it would be better to speak of restoration of the old situation. The economic dominance of Europe and North America is a product of the Industrial Revolution and thus not older than two hundred years. Before that time, the distribution of wealth and poverty around the globe was more equal. It has been calculated that until the beginning of the 19th century China and India used to account for more than half of the world gross product.[6]

Of course, impressive growth of the GNP of a country does not mean that all of its citizens profited equally. When a third world country becomes involved in the global economy because of its low wages, the differences of income in that country can increase substantially, because a small upper class may keep most of the profits in its own pockets. However, judging by similar situations in the past, it is to be expected that in the long run a larger part of the growth of national wealth will find its way to the lower classes. In the

6 A. Maddison, *Contours of the world economy 1-2030 AD* (Oxford University Press, 2007).

early years of the Industrial Revolution in Britain, the working class lived under miserable conditions while factory owners were gathering enormous riches, but one or two generations later the circumstances of the proletariat improved considerably, towards a higher level than they had known before the Industrial Revolution. In the same way is to be expected that globalization will ultimately spread material wealth more evenly over the world than is the case today.

The poorest of the poor in Asian countries that are strongly involved in economic globalization have already since many years seen their living standards improve. The income of the poorest fifth of Malaysian citizens has on average grown by 5 or 6% per year. The poorest fifth of the Chinese population experienced an annual growth of income of nearly 4% in a time when the national economy grew at a pace of 8 to 12% per year. So, everybody has profited, but those who needed it most have until now profited least.

Broader proof of improvement for the poorest can be found in a report issued by the World Bank in 2012. It shows that between 2005 and 2008 the number of people living on less than one euro a day has for the first time in history decreased in every country of the world. The speed of improvement is so fast that the Millennium Goal of the United Nations to reduce extreme poverty by half was already reached in 2012, three years before the deadline that had been set.

In the past there have been periods with other kinds of global economic relationships than today's popular model of work sharing between highly developed countries and countries with low wages. Such a bipolar system existed in the time of colonialism as well, but then the two parties involved were the colonies as suppliers of raw materials and agricultural goods and the colonizing industrial countries as consumers. In medieval Europe things were done quite differently. Hanseatic trade agreements created an intricate European network of specialized suppliers, traders and buyers of end products, all of whom acted as partners of more or less equal standing. The roles of different countries and groups of countries in the global economic network may in the future shift towards this old Hanse model, away from the bipolar model.

In any case, it seems certain that global interdependence will grow further, unless a new world war breaks out or some ideological conflict leads once more to a situation of cold war with trade barriers and boycotts. If no political catastrophes happen, the world could within a few decades merge into a single economic system. In the second half of the 21st century mankind would then experience both the stabilization of size of the world population and economic unification.

The role of multinational industrial corporations and financial institutions in the process of economic globalization is large. They control a big part of global investment capital and nearly three quarters of world trade. Their decisions direct trade streams, determine where branches of industry will grow or decay, et cetera. This means that many choices of strategic importance to the world economy are made outside the reach of governments, parliaments or other democratic institutions. Politicians should consider it their task to set up some procedure for examining such important business decisions with respect to their effects on public welfare before granting permission to execute them. This should happen at a supranational level, preferably even by a body with global authority, but the reality is that the necessary political and legal mechanisms are still lacking.

In a later chapter we will consider the question whether the earth could in the course of the 21st century become politically unified. At this point I only note that peaceful co-existence of nations and economic globalization constitute a textbook example of positive feedback. (One speaks of positive feedback between A and B if more A leads to more B which in turn leads to more A, et cetera.) As international conflicts diminish, more room exists for economic integration, and increased integration calls for further peace-promoting measures. If one country has large business interests in another country, then it knows that it will harm itself if it starts hostilities against that country. In short, globalization forces governments that act rationally to avoid war.

Until now, events in Ukraine are an example of this mechanism. The interwoven interests of Russia and the European Union (gas, technology, loans) so far prevent the outbreak of full-scale war. The situation was different in the Balkans after the disintegration of Yugoslavia. The people of Yugoslavia had the bad luck that no vital economical interests of third parties were at stake. Of course, the situation could quickly become very ugly if Russia would decide not to act rationally.

Economic integration brings many advantages, but it has the drawback that slumps and booms of the world economy will tend to be more extreme. When economies of different parts of the world are only loosely coupled, the booming economy of one region can pull another region out of recession. This mechanism doesn't work anymore if the world economy behaves as a single system. Given that the trend of globalization cannot be stopped, one would have to invent mechanisms which can automatically dampen ups and downs of the economy. In order to accomplish such a feat, one would first have to transform economics into a real science, which may be unrealistically ambitious.

Cultural globalization

Cultural globalization is the spreading of ideas, customs, ethical values, knowledge and patterns of behaviour without regard for national borders. It happens automatically as soon as we interact with people from elsewhere, visit foreign countries or learn through the media what people think and do in faraway places. It becomes accelerated by the same technological developments that speed up economic globalization: cheap and fast transportation, telecommunication and digitization of information.

Among specialized, highly educated and highly qualified professionals, cultural globalization is the natural state of affairs. In the worlds of science and art somebody's nationality has seldom been considered relevant. In these fields one is since long used to world-wide migration, cooperation and competition. The same has in our time become true in other circles, like top-class sport, crime syndicates and board rooms of corporations. It is a trend that is spreading further, first of all to highly-skilled people in technical professions.

Everybody experiences cultural globalization in the fields of music, sports and cuisine. Growing closer together in such aspects of everyday life consists to a large extent of almost epidemic spreading of Western (American) consumer habits. However, it is not necessary to fear that the whole world will by the year 2050 be 'Coca Colonized'. As the economic weight of other parts of the world grows, so will their influence in other aspects of life. History shows that cultural impact of one country on others mirrors with some delay its political, economic and military power. In this way, Europe has known periods of great cultural influence from Spain, France, Austria (the Habsburg empire), The Netherlands, England and Germany, but never to a degree of predominance that would have smothered other cultures completely. There has always been mixing, adaptation and addition of new elements – that is, change and renewal. In the future it will be similar on a global scale. The present economic rise of China, India and South America will soon be followed by an increased influence on ways of life elsewhere on earth, and at some later time it will be the turn of other countries and regions.

The Netherlands are an example of the fact that a state with a strong economy cannot simply overpower its weaker neighbours culturally, even if they remain closely linked for centuries. In the Middle Ages the counties of Holland and Zeeland, the bishopric of Utrecht, the duchy of Gelre and the provinces of

Friesland, Groningen and Overijssel waged several wars among each other, but in the 16th century they were forced to unite into the Republic of the Seven United Netherlands. Holland has always been dominant in this federation. The long cohabitation has of course made regional peculiarities less pronounced, but people in the weaker provinces certainly did not adopt the customs of Holland to the point of losing their own identities. One could even say that Holland has been the province with the largest loss of identity. Because the other provinces imported quite a lot of the culture of Holland, relatively little is still recognized as typical of Holland. The other provinces exported fewer of their customs, with the result that their surviving customs and traditions are clearly recognizable as their own cultural heritage.

Although cultural globalization will more often be enrichment than impoverishment, it will always provoke resistance. The more cultural mixing becomes a reality, the louder the call to build defences against foreign influences will sound. It is a rear guard battle, but xenophobes and primitive nationalists can still make a lot of trouble. Other critics, known as anti-globalists, maintain that all forms of globalization are nothing but new manifestations of colonization and imperialism. They do not wish to recognize that poor countries decide themselves whether or not to play a part in economic globalization, and that countries that did join, generally fared better than those that didn't. In reality, those critics are themselves behaving as old-style paternalistic Westerners, claiming to know best what is good for others.

Climate change

The greenhouse effect

The economic growth after the Second World War, since the eighties of the last century further accelerated by the process of globalization, has immensely increased the use of natural resources, especially of fossil fuels. That does not mean that there is an acute danger of supplies of coal, oil and gas running out. There is still coal for many centuries, even in the highly unrealistic case that the world would continue to burn coal at the same rate as today. The known reserves of natural gas are also enormous. About half of the technically and economically recoverable oil reserves of the earth have already been used up, but it would be possible to produce synthetic oil from coal and gas at a price that is not terribly high.

So, there is no immediate shortage of fossil fuels, but as everybody knows nowadays, that is not the point. The problem is that the burning of enormous quantities of fossil fuels has strongly increased the concentration of carbon dioxide (CO_2) in the atmosphere. This is alarming because CO_2 is a greenhouse gas. Its accumulation in the atmosphere causes the temperature of the biosphere – land and sea surfaces and the bottom layer of the atmosphere – to rise.

How does this work? The balance between the incoming light from the sun and the heat radiation that earth emits into space, determines whether the temperature at the surface of the earth rises, falls or remains unchanged. Sunlight is most intense at visible wavelengths. (That is of course not just luck. The eye has in the course of evolution adapted itself to the spectral distribution of sunlight.) Visible light is electromagnetic radiation of quite high frequency. Greenhouse gases are transparent to it. Heat radiation is also electromagnetic, but in the infrared region. Those very low frequencies are strongly absorbed by CO_2 and other greenhouse gases. The absorbed infrared radiation is immediately re-emitted, but that happens in all directions, not just into space. The result is that a large part of the heat radiation comes back into the biosphere

However strong or weak the greenhouse effect is, the atmosphere will settle into a situation of equilibrium between incoming and outgoing radiation. The average temperature will become such that the fraction of heat radiation escaping into space carries away the same amount of energy as is brought to earth by sun light. If the escaping fraction of heat radiation is small, then the primary amount of that radiation must be large (that is, the temperature must be high) in order to maintain equilibrium with the incoming sunlight. In short, more greenhouse gas means a higher temperature.

One must understand that the worries about CO_2 in the atmosphere purely concern the fast rate at which the concentration is rising. It is not that greenhouse gases would be evil in principle. On the contrary, we cannot do without them. Thanks to a strong natural greenhouse effect in our atmosphere, the average surface temperature of planet earth is about 14 °C, comfortably within the temperature range where water is a fluid. Without greenhouse protection, earth would be a freezer set at -18 °C . That is an effect of 32 degrees, while the discussion about warming of the earth due to human activity centres on the question whether we can keep that limited to two degrees. The effect of 32 degrees is mainly due to the enormous amount of water vapour in the atmosphere. Water (H_2O) and all other gas molecules consisting of three of more atoms can in the same way as CO_2 absorb and re-emit infrared radiation. The contribution of water vapour to the natural greenhouse effect is three or four times larger than that of CO_2. Methane (CH_4) contributes two or three times less than CO_2.

Through world-wide industrialization the concentrations of CO_2 and CH_4 have risen by approximately 40% and 150% during the 20th century. The result is that the present greenhouse effect is stronger than it has been at any moment in the last 800,000 years – which is the length of time for which reliable data are available from ice core drillings.

As long as the concentrations keep rising, the mean temperature will rise as well. Reduction of the emissions of greenhouse gases to almost zero would *not* quickly make earth cooler again. The methane concentration would within a few decades fall back to its pre-industrial level, but the amount of CO_2 in the atmosphere would only very slowly go down, because this gas is chemically quite inactive. It could take at least a century before half of the atmospheric CO_2 due to human activity would be absorbed by the oceans or used up by plants in the process of photosynthesis. In the end, the excess of CO_2 in the atmosphere would mostly be taken up by the oceans, but that would be a very slow process. After one thousand years about 20% of the original emissions would still be in the air.

So, on a human time scale the warming of the biosphere is not just a passing event that could be fixed quickly by stopping emissions. On the other hand, compared to processes playing on geological time scales, the greenhouse effect caused by mankind is a very minor occurrence, both in its duration and in its effects. The earth has during long periods of its history been much warmer than today. We live in a cold period: since more than two million years the planet is in an era of ice ages. To be more precise, we happen to live in a pause of several tens of thousands of years between two ice ages.

Could the next ice age start soon, so that we could happily continue burning fossil fuels without having to worry about warming up the planet? The answer is a definite no.

For the advent of an ice age, the subtle long-term periodic changes in the orbit of the earth around the sun must be in such a phase that the amount of sunlight reaching the earth is minimal, and at the same time the concentration of greenhouse gases must be low. At the present orbital constellation, the atmosphere would have to be extremely clean of greenhouse gases, even cleaner than before the Industrial Revolution.

If our children or grandchildren will want to return to air temperatures of, for example, the middle of the 20*th* century, they will have to intervene actively, either by extracting enormous amounts of greenhouse gas from the atmosphere, or by shielding the earth from incoming sunlight. I will return to the question whether such actions would be wise, or whether it would be better to make the best of living on a warmer earth.

How much global warming at what price

Between 1900 and 2010 the average temperature at the surface of the earth (land and sea) has risen by almost one degree centigrade. Most of the increase occurred after 1970. Impressed by that fast warming, nearly all countries of the world in December 2009 signed a declaration at the United Nations Climate Change Conference in Copenhagen to the effect that they 'recognize the scientific view that the increase in global temperature should be below 2 degrees Celsius'. In other words, after the rise of around one degree with respect to the pre-industrial level in the course of the 20*th* century, any further warming should remain below one extra degree. This ambitious goal was translated into an upper limit on the concentration of CO_2 in the atmosphere. Consensus had, however, not been reached on binding agreements regarding the reduction of emissions, so the declaration stayed mute on that crucial issue.

In the period 1970-2010 energy consumption approximately doubled. Since the amount of CO_2 emitted per unit of generated energy dropped only by ten percent in the course of those forty years, the emissions of CO_2 also nearly doubled. The Copenhagen agreement has not yet brought big changes. In 2012 the world energy production still relied for 87% on coal, oil and gas.

More encouraging is the trend in energy efficiency. Energy consumption per unit product has dropped globally by about 40% between 1970 and 2010. The decrease is mainly due to a shift in economic activity from industry to services, from chimney to office. This trend will probably continue at about the same rate in the next decades.

The volume of goods and services produced will approximately follow the growth of the world population. Taking into account that globalization will lead to some wealth increase for the average citizen, the production curve may even rise a bit steeper than the population curve. Fortunately, the world population will from the middle of the 21st century hardly increase anymore. After that, the gross world product could only grow further if living standards keep improving, but that is not likely to be a big effect. It may count less than steady increases in energy efficiency. This would mean that energy use will around the middle of the 21st century reach its peak and then will start to decline slowly.

What does this scenario of future energy demand – increase about parallel to world population growth until about 2050 and then a slow decrease – mean for greenhouse gas emission? In order to answer this question, we ought to know how fast the share of fossil fuels in energy production will fall, but this is very unclear. It is a matter of political choices, and thus of electoral calculations and horse trading.

The International Energy Agency (IEA) has estimated that the share of fossil fuels in 2050 will be around 60 or 70%. A share of 60% in the energy production of the year 2050 is in absolute terms roughly equal to the 87% share realized in 2012. In other words, CO_2 emissions will still be at the same level as today, which means that the concentration in the atmosphere will in the coming decades keep rising at the same rate as today.

The Intergovernmental Panel on Climate Change (IPCC)[7] has in its *5th Assessment Report*, published in 2013-2014, presented estimates of the rise of the average surface temperature during the 21st century for three different scenarios.[8] If the emission of CO_2 continues in the coming century at the present level, the temperature in the year 2100 will be approximately 4 °C higher than in 2000. In the improbable case that emissions will within a few years be drastically reduced, the temperature will rise by just one degree. In the most realistic scenario, which assumes that fossil fuels will still remain our main source of energy for several decades, the temperature in 2100 would be two degrees higher than today. That is one degree less than the estimate by the International Energy Agency in its *World Energy Outlook 2013*.

How bad is it if the earth indeed warms up by another two or three degrees, on top of the one degree temperature rise of the 20th century? For the planet earth it would not matter much. It has been far warmer in the

7 The IPCC is a scientific body operating under the auspices of the United Nations. Its mission is to evaluate scientific information relevant to climate change. The IPCC received the 2007 Nobel Peace Prize. All IPCC reports are available online at www.ipcc.ch.

8 The scenarios correspond to CO_2 concentrations in the year 2100 of about 420 ppm, 600 ppm and 940 ppm. For comparison: around 1750 the concentration was 278 pm and in 2013 the value of 400 ppm was touched. One ppm (part per million) corresponds to a concentration of 10^{-6}.

past. For many life forms, including man, it is a different story, not so much because of the temperature rise in itself but because of the great speed at which the change happens.

In the second half of the 21st century the warming can have consequences that are much more serious than problems that may in the first decades of the century arise from population growth and globalization. To start with, the sea level will rise. Hundreds of millions of people live in low coastal regions, especially in East and South Asia (China, India, Bangladesh, Vietnam, Indonesia, Japan, Thailand, The Philippines). The predictions in the *5th Assessment Report* of the IPCC for the end of the 21st century vary from 40 cm for the case that during this century only small amounts of CO_2 will still be emitted to 70 cm if the world continues to burn fossil fuels at the same rate as today. In the latter case the sea level would by the year 2300 have risen by several meters.

The main reason for the rise of the sea is that water expands when it becomes warmer. The contribution of melting land ice is in the beginning small, but on the time scale of centuries or millennia a partial melt of the ice cap of Greenland (and maybe of glaciers on the west coast of Antarctica) could lead to an extra rise of the order of five or seven meters. According to the IPCC Report there is a certain threshold value of global warming above which the Greenland ice would almost entirely disappear in the course of a thousand years or longer. It is alarming that this threshold would according to model studies already be passed if it becomes three degrees warmer than in 2010.

This doom scenario could be a reason not to reconcile oneself with a warmer earth and to start efforts to cool it down again. Glaciers don't care about a century more or less, so there is still ample time to think of a realistic cooling method. I will soon return to this question, but let me say now already that I doubt if man will ever really try a forced cool-down. Because the rise of the sea level due to melting of land ice would progress only very slowly, each generation of politicians and government officials will be tempted to pass the hot potato of costly measures to the next generation. One would rather limit activities to increasing the height of a dam or dike now and then, or sometimes relinquish a strip of coastal land.

Climate models all agree that warming will be largest in polar regions, and the facts already show it. In the 17*th* century every expedition trying to find a northern sea route from Europe to Asia got stuck in the ice, but at present the route is practicable in summer time. Roughly speaking, warming will be least noticeable in equatorial regions, although the details of physical geography can lead to considerable variation from region to region. In general, ecosystems will shift away from the equator towards the poles.

Patterns and amounts of rain fall will change. Climatologists predict a wetter Sahara and a shift of subtropical deserts into higher latitudes as a consequence of enlargement of the so-called Hadley cells, the regions of atmospheric circulation between the equator and the subtropics. Around the equator warm air rises to 10 or 15 kilometres, releases rain as it flows towards the subtropics, descends as a very dry fall wind and flows close to the earth surface back to the equator.

Climate models indicate that the Hadley cells will expand with rising temperatures towards the poles, with the result that the desert zones of the northern and southern hemispheres will in the 21st century shift by two or three degrees of latitude towards the poles.

In a warmer atmosphere the water cycle of evaporation, condensation and precipitation will be sped up. This is just one aspect of the fact that the energy content of the atmosphere will increase, which will generally make the weather more turbulent: more rain fall, more and heavier storms, thunderstorms and hurricanes.

Most worrisome is the possibility that loops of positive feedback (circles of cause and effect that boost each other) may arise. It has for instance been suggested that the melting of the vast regions of permafrost in Siberia and Canada will release enormous amounts of CH_4 and CO_2, which will further increase the greenhouse effect, causing the temperature to rise further, so that the tundra will melt even faster, et cetera, until all permafrost is gone. It is as yet unclear if this specific example of a possible feedback loop is a realistic danger. Up to now there is no sign that melting permafrost has significantly increased the concentration of greenhouse gases in the atmosphere. It has even be claimed that the thicker vegetation on the thawed tundra takes up at least as much extra CO_2 from the atmosphere as is released when the permafrost melts.

A feedback which certainly is real, is that the melting of sea ice in the Arctic will increase the speed at which the earth is warming, for the simple reason that ice reflects 70% of incoming sunlight into space, whereas open water absorbs 94% of the sunlight.

The oceans will in due course absorb enormous quantities of CO_2 from the atmosphere. That starts a chain of chemical reactions, which in the end results in an increased acidity of the water. This means that the water will contain more free hydrogen ions, that will tend to combine with carbonate ions to become bicarbonate. The decreased carbonate concentration is a very serious matter, because many micro-organisms at the base of the marine food chain need carbonates. The speed at which the oceans are becoming more acid is at present many times faster than ever before in the last 300 million years. In that long period there have been four episodes of mass extinction of life forms in which a high degree of acidity demonstrably played a role. Acidification of the oceans has been called the equally evil twin of the warming of the biosphere.

Is it possible to quantify the economic effects of global warming? Disasters like the dying of coral reefs or the disruption of arctic ecosystems can of course not be expressed in sums of money, and a truly excessive warming of the biosphere would result in catastrophes that one cannot put price tags on, but if we restrict ourselves to a moderate warming of say two degrees between the years 2000 en 2100, and only consider matters suitable for pricing, of which order of magnitude would the economic damage be? The IPCC has dared to investigate this question in its *5th Assessment Report*. It was estimated that the loss of annual income as a consequence of moderate warming would for the average world citizen lie somewhere between 0.2 % and 2%.

This requires two remarks. First, the range 0.2% - 2% only indicates the uncertainty in the calculation of the *average* effect on income. It does not say anything about how the effect may differ from country to country or from region to region. Those differences may be very big. The gross national product of a country suffering from desertification may be halved, while another country becomes wetter and flourishes as never before. Second, there will be regions on earth where the economic consequences of other developments – civil war, a strong shift in size or composition of the population, discovery of valuable minerals, whatever – are so large that they mask the effects of climate change.

Waiting for governments to act

When will governments start to take serious action to prevent a climate crisis? Which of the three scenarios of the IPCC report will become reality? The most optimistic one, which assumes that there will be no large emissions of greenhouse gases after the first decade of the 21*st* century, is obviously unrealistic. Switching on a large scale from fossil fuels to cleaner sources of energy is an enormous task and requires gigantic investments. That process will take decades. The darkest scenario – continued burning of fossil fuels at undiminished rate – will become reality if the biggest polluters, the United States and China, stay with their present behaviour. It is to be feared that these countries will only changes their ways when their own citizens start to suffer the consequences of climate changes, physically, economically or in any other way, but in any case in such a manner that the relationship between cause and effect is undeniable.

Even if hard lessons will make the greatest polluters repent within the next ten years, it will not be possible to replace most of the coal and gas plants by clean power sources before the middle of the century. Only late in the 21st century could one reach the stage where power generation does not lead to any significant release of CO_2. Such a course of events agrees more or less with the middle scenario of the IPCC report. For this case a further warming by about two degrees during the next hundred years has been estimated, and the sea level is expected to rise by about half a meter.

That may not sound catastrophic, but in reality it will cost a lot of effort and money to absorb the consequences of such a seemingly small change, on top of the sums one will by that time already have invested in clean energy production. Governments will have to be courageous and energetic, else they will give in to the temptation of passing problems on to their successors. Climate change is happening extremely fast compared to the time scales of biological evolution and geological processes, but compared to the period of service of any government official, everything related to the climate occurs at a snail's pace.

A sea level rise of half a meter or maybe one meter will require huge engineering projects in water management in Bangladesh, Florida and many other coastal regions, but if that takes place over a period of a century, the water will rise by only a few centimetres during a term of office of the governor of Florida. For the average voter that is unnoticeable, and thus unimportant, so the governor will not want to spoil his chance of re-election by introduction of a tax to pay for proper defences against the rising sea. Governments and officials elsewhere will for similar reasons prefer not to react to desertification or the transformation of the melting tundra of Canada and Siberia into endless swamps.

Wherever politicians depend on the favour of voters, so in all democracies, it will be very tempting the go along with the short-sightedness of the electorate and choose stopgap measures. It will not be better in an autocratic system of government, where the Will of the Leader is the law. Just considering the rules of probability, so without any negative assumption about the capacity of judgment of autocrats, it is clear that any initially chosen course of action has very little chance of reaching its goal if adjustments along the way are forbidden, as is usually the case under an Infallible Leader. The winding path followed in a democratic system will generally get closer to an acceptable solution.

Loss of land to the rising sea, appearance of a new desert here and greening of a former desert there, subarctic regions becoming habitable, all kinds of changes in the physical geography will ultimately lead to the migration of hundreds of millions of people, but this too will happen quite slowly. One should not imagine hordes of fugitives chased by natural disasters, civil wars and epidemics. It could somewhat resemble the present flows of people from Africa to Europe and from Latin America to the United States, but on a bigger scale and for that reason probably based on agreements between the nations concerned, instead of today's practice of illegal entry..

Attempts by authorities to make the adaptation of humanity to a warmer environment proceed in a more or less orderly fashion will require a large amount of international coordination. For instance, a system of immigration quota in countries and regions that profit from the climate change is to be expected. The optimist may hope that the warming of the earth will in this way, through the force of circumstance, bring about a habit of international cooperation. The pessimist (realist?) anticipates a chaotic time of colliding interests, alliances of nations harmed by climate change against nations that profited, economic warfare, violent border conflicts and more.

If mankind indeed succeeds in reducing the emission of greenhouse gases in the course of the 21*st* century to a negligible amount, then the warming will end, but the biosphere will still stay warm for centuries to come. Left to nature, cooling will be a very slow process. Would it be wise to force a quicker cooldown? The answer is yes if the temperature has become so high that the ice cap of Greenland is melting away. If one would allow that to continue unchecked, the seas would in the course of centuries rise by many meters. If at all possible, it would be worth enormous costs to stop that. I will discuss methods of forced cooling later.

If there is no threat of large-scale melting of land ice, it seems wiser to accept the warming as a fait accompli. Attempts at forced cooling are not guaranteed to work, may have unpleasant side effects and will meet understandable opposition from states that profited from the temperature rise. Nations along the greening southern rim of the Sahara would not like the prospect of their land becoming a desert again.

In any case, even if the warming indeed will come to a halt around the end of the 21st century, the aftermath will presumably dominate international politics until far into the first half of the 22nd century, unless a different kind of global crisis with even worse consequences occurs.

Towards which clean energy?

Which form of clean energy production will replace fossil fuel plants? Will many nations dare to put their money exclusively on the politically correct windmills and solar cells, in spite of the serious shortcomings that the first will not supply energy in calm weather and the second not at night? In order to have a guaranteed supply of energy at any time, these highly variable sources would have to be backed up by a gigantic buffer system. That is imaginable, but would be very impractical. One could for instance build water reservoirs above and inside empty mines or at the top and foot of mountains, pump water up at times of overcapacity and let it stream down to drive a turbine at night or during calms. Or one could use a surplus of electricity to make hydrogen gas from water and later use the hydrogen as a power source.

One could think of many other ways of realizing energy reservoirs, but it will always remain true that a system purely based on energy sources with pronounced highs and lows must be very complex and extremely expensive, because of the double facilities for production and for buffering. So, a realistic system of power generation without any emission of greenhouse gases cannot be built on wind and sunlight alone. One needs an equally 'green' power source that is guaranteed to be continuously available, either in addition to windmills and solar power, or as the stand-alone solution. Such a technology exists. It is called nuclear power generation.

Apart from its intrinsic pros and cons, there is of course the problem that nuclear energy is in most Western countries presently not popular with the electorate. I will discuss facts and feelings about nuclear power in a later paragraph, but the point I want to make here, is that a system of power generation without any production of greenhouse gases will not be possible without a nuclear component. If a country rejects nuclear energy, as Germany has done, wind and solar power will have to be backed up by conventional power stations fuelled with coal, gas or oil.

The German decision was the result of a promise made hastily in election time. German voters, shocked by the disaster at the nuclear power plant of Fukushima, had loudly made it clear that they wanted existing nuclear reactors to be shut down and no new ones ever to be built on German soil. The political parties listened, with the result that nuclear power stations will indeed be shut down prematurely. It is nice that at the same time huge investments are being made in wind parks, but the fact remains that Germany is *not* on the road towards elimination of CO_2 emissions. Now and then there will be days or even weeks without much wind, so a full array of power stations burning fossil fuels has to be kept operational. Since these are only meant as backups, their exploitation will not be commercially attractive, which means that power companies will demand huge subsidies to maintain and operate the backup system.

Quite a few other Western countries also appear to be heading for the German model of wind and sun plus fossil fuel stations. The latter will in most cases burn gas, because a gas power plant can relatively easily switch back and forth between different levels of output, which is an essential requirement if the plant has to come into full action at each calm. Being dependent on natural gas is of course risky if a country binds itself to deliveries from unpredictable, unstable Russia. This is the second fault in the German decision to drop nuclear energy.

In the field of energy production France has shown itself wiser than Germany. Since many years it generates about 75% of its electricity with nuclear reactors. If the rest of the world had followed that example, much less greenhouse gases would have been produced in the past decades and the present threat of an acute climate crisis would have been spared us. France will presumably go entirely for nuclear energy, maybe adding a symbolic windmill park in the Bretagne and a solar power park in the Provence.

Although each European country is still planning its own energy policy, reality will soon be that all national choices are mixed into one hotchpotch. The electricity grids of many neighbouring countries are already connected to each other and the rest of the EU is to follow soon. The idea is to create with this infrastructure a single European energy market. Germany and other countries with large investments in the variable energy sources wind and sunlight may in the long run prefer to fill their periodic shortages with electricity imports from countries that chose for nuclear energy. That will be much cheaper than maintaining a chain of gas powered stations inside the own country. In this way Germany would indirectly return to the use of nuclear energy.

Summarizing: in two or three decades from now, the clean but unreliable energy sources wind and sunlight, reliable and clean nuclear reactors and reliable but dirty gas powered plants will be the mainstays of the world power system. Gas plants may disappear when nuclear energy loses its bad name after many years of safe operation of the latest generation of nuclear reactors. Horizon polluting wind mills might then also fall into disgrace. The result would be that around the middle of the 21st century most of the world is powered by nuclear energy.

Energy production by nuclear fission is a clean technology. No greenhouse gases are produced and the tiny volume of nuclear waste can easily be stored under ground, isolated from the biosphere. Worldwide, about 13% of electricity was in 2010 generated in nuclear reactors. In the European Union it was 30%.[9] The fuel uranium is as plentiful as tin. The reserves that can easily be mined are sufficient for at least 250 years at the present level of consumption. (By recycling used uranium that time could even be doubled , but lack of nuclear fuel will anyhow never be an issue, because *fission* is likely to be replaced by nuclear *fusion* before the end of the 21st century. More about that follows below.)

9 International Energy Agency, *Key world energy statistics 2012*; www.iea.org/publications/freepublication/ kwes.pdf.

Some comparative studies have been published about the health risks and the chances of severe accidents related to power generation with fossil fuels, hydropower and nuclear fission. In each case the entire chain of activities was investigated, from fuel mining and transport (mining accidents, pipeline sabotage, ...) through operation of the power station (radiation leaks, collapsing dams, ...) up to electricity distribution to the customers. It is no surprise that air pollution from the burning of coal, oil and natural gas turned out to be the largest health risk.[10] Also on the subject of deadly victims per unit of generated power the fossil fuels did worse than nuclear fission and hydropower, although the good score of waterpower only applies to highly developed countries.[11]

These good marks for nuclear energy contrast starkly with the image of nuclear power in the eyes of the general public, which has been formed by the spectacular accidents at Three Miles Island in 1979, Chernobyl in 1986 and Fukushima in 2011. These disasters indeed caused very severe material damages, but measured by numbers of victims they must be classified as tiny incidents compared to the numbers of lives lost in coal mines and at oil and gas wells. The preference in a country like Germany for dirty and dangerous fossil fuels over clean and relatively safe nuclear fission is irrational, but for that very reason difficult to correct.

10 A. Markandya and P. Wilkinson, 'Electricity generation and health', *Lancet* 370 (2007), p. 979-990. Online in PubMed (www.ncbi.nlm.nih.gov/pubmed) under nr. 17876910.

11 P. Burgherr and S. Hirschberg, 'A comparative analysis of accident risks in fossil, hydro and nuclear energy chains', *Human and ecological risk assessment* 14 (2008), p. 947-973. Online in http: //gabe.web.psi.ch/pdfs/ _2012_LEA_Audit/TA01.pdf.

The instinctive fear of the incomprehensible magic of nuclear energy will be a severe obstacle when the world tries to switch from dirty to clean forms of energy production. Choosing for unreliable wind mills backed by fossil fuel plants rather than reliable, clean nuclear power cannot be defended with rational arguments. The three severe nuclear accidents mentioned above happened in reactors designed in the fifties and sixties of the last century. A modern reactor will always shut itself down before a meltdown can occur, even if operators make grave mistakes (Chernobyl) and even if forces from outside destroy parts of the infrastructure (Fukushima).

Most voters – and consequently most politicians – will probably not overcome their deep aversion against nuclear energy until it becomes clear that the German system of 'free-range power' must frequently be supported by nuclear power imported from France, but in several quarters the insight is already now gaining ground that a more rational attitude towards nuclear energy is necessary. Among those who promote this message is a group of people calling themselves 'Environmentalists for Nuclear Energy' (www.ecolo.org). They call opposition against nuclear energy by people who are concerned about the environment 'the greatest misunderstanding and mistake of the century'. In 2004 James Lovelock, father of the Gaia hypothesis, created a media sensation when he broke with his fellow activists for a greener world by declaring that only nuclear energy could still stop further warming of the earth. The IPCC has supported the case for nuclear energy as a means of gaining control over the climate crisis in its *5th Assessment report.*[12]

12 In the chapter *Energy systems* of the report by Working Group II.

Power production through controlled nuclear *fusion* – the fusion of light atomic nuclei at extreme high temperature and density – is since 60 years a promise that will become reality within 25 years. It is in principle an ideal technology. The fuel for a fusion reactor consists of isotopes of hydrogen or other light elements that will never be in short supply, the danger of a meltdown does not exist and no long-lived radioactive waste is produced. And on the side of sentiments: everybody should love nuclear fusion, also and especially ideologically inspired champions of green energy, for it is the process that makes the sun shine. In other words: nuclear fusion is the mother of all natural resources. There can be nothing better than directly tapping the mother energy of nature.

The physics of nuclear fusion is simple and well understood, but its realization under controlled circumstances is technologically extremely difficult. Fusion provides energy because the sum of the masses of two light atomic nuclei merging together to form a single, larger nucleus, is larger than the mass of that single nucleus. The missing mass appears as pure energy according to Einstein's formula E(nergy) = m(ass) x c^2 (speed of light squared). The most advantageous reaction is the fusion of deuterium ('heavy hydrogen') and tritium (an even heavier type of hydrogen) into helium plus a neutron (a free nuclear particle). Deuterium occurs in enormous quantities: 0.015% of all water molecules contain deuterium instead of ordinary hydrogen. Tritium hardly occurs in nature, but can be easily made by irradiation of the element lithium with neutrons. It is a radioactive isotope with a half life of twelve years, but its radiation has so little energy that it cannot even penetrate our skin. Tritium is only dangerous if one swallows a large amount of it.

Nuclear fusion is not a chain reaction. That makes it an inherently safe process. The technical problem is that fusion only proceeds under conditions of temperature and pressure that occur deep inside stars. In order to obtain an acceptable energy yield, a fusion reactor must operate at temperatures around one hundred million degrees centigrade. That amounts to continuously creating similar conditions as inside a hydrogen bomb and at the same time withdrawing energy from that inferno at a controlled rate.

Strong efforts to development a fusion reactor are still under way. The largest and most promising project is ITER (International Thermonuclear Experimental Reactor),[13] at the moment under construction at a cost of 10 billion euro at Cadarache in the south of France. It is a joint project of the European Union, the United States, China, Japan, South Korea, Russia and India. In ITER a plasma of deuterium and tritium will be confined and compressed by magnetic fields and heated to 150 million degrees centigrade by electric currents through the plasma. The principle has been tested on a smaller scale in the Joint European Torus in Oxford. Already in 1997 an experiment was performed there in which 16 megawatt of fusion power was released during a short time. One expects the first fusion reactions in ITER around 2030. The aim is to finally generate 500 megawatt of fusion power, but only during periods of at most 15 minutes at a time. The fact is that it is a research facility, not a prototype of a power station.

13 Website www.iter.org.

There are competing ideas. At several places experiments are being done or planned to test other methods to realize controlled fusion. Best known are the implosion experiments at the National Ignition Facility of the USA.[14] Here a hollow globule of which the insides are coated with a frozen mixture of deuterium and tritium is irradiated with an extremely intense laser pulse. Explosive evaporation of the outer wall of the globule should press the deuterium and tritium nuclei so strongly together that they fuse. In 2013 it was demonstrated that the idea in principle works – some fusion was observed – but there is a long way to go before the fusion energy produced would even equal the energy used for the laser ignition. In addition, it is far from clear how to transform this technique into a usable power station.

It is very likely that the first actual fusion plants will be siblings of ITER. People working in this field expect the first one to become operational around the middle of the 21*st* century. This is maybe too optimistic, as is often the case when experts talk about their own work, but even if it takes a quarter of a century longer before fusion plants start to deliver electricity, the advantages will be so great that nuclear fusion will very quickly supersede nuclear fission and the various 'free-range' sources of energy (windmills, solar cells, hydroelectric, tidal energy, …).

In summary: I expect that energy production from fossil fuels will in the course of the 21*st* century first be replaced by nuclear fission and free-range electricity, and that this array of energy sources will late in the 21*st* century give way to nuclear fusion. Free-range

14 Website lasers.llnl.gov.

electricity has a long-term future only in thinly populated areas, where local energy production is cheaper and simpler than constructing an elaborate distribution network from a single big power station.

Climate manipulation as an emergency measure

Once the warming up of the biosphere has *stopped*, it would not be wise to make attempts at forced cooling, except if it were found that all arctic land ice would otherwise melt away over the centuries. It is a different question whether climate manipulation might make sense *during* warming. Suppose governments conclude that they cannot sufficiently fast come to grips with the heart of the problem – excessive emission of greenhouse gases – to prevent a catastrophic temperature rise. Could one resort to some kind of emergency measures that would temporarily slow down the warming, just to buy time for resolving the basic problem of high emissions? Such a situation could really arise, given that the global volume of emissions is still growing. In spite of the economic dip after the banking crisis of 2008 and in spite of the earnest efforts by some countries to switch to cleaner sources of energy, the emission of greenhouse gases has between 2000 and 2010 risen by 2.2% per year. That is considerably faster than in the period 1970-2000, when the average annual increase was 1.3%.

If one should conclude, in twenty or thirty years from now, that the biosphere keeps warming up much too fast, then one could think of two kinds of stopgap measures: decreasing the amount of sunlight absorbed by the atmosphere and the surface of the earth, or artificially removing greenhouse gases from the atmosphere. Because of the risk of unforeseen and uncontrollable side effects, such steps should only be considered if he climate *crisis* is becoming a climate *disaster*.

Decreasing the amount of absorbed sunlight does not tackle the primary problem, the high concentration of CO_2 in the atmosphere, but it has the advantage that it can already within a few years have a noticeable effect on the temperature. One could stimulate cloud coverage and use sprays to increase the reflective power of clouds. Or one could introduce a heavy concentration of aerosols (tiny liquid or solid particles) into the stratosphere in order to scatter sunlight back into space.

The most realistic proposal in this category may be the idea of the Dutch Nobel laureate Paul Crutzen to increase the reflectivity of the earth by injecting sulphur into the stratosphere.[15] Until a few years ago large amounts of sulphur were being released into the atmosphere in the form of sulphur oxide, which is a product of the burning of fossil fuels. It had a favourable effect on the warming of the atmosphere. Crutzen estimated that it neutralized 25% to 65% of the greenhouse effect of CO_2. Most developed countries have nowadays banned the emission of sulphur oxide because it is a health hazard. Crutzen has proposed that one could resume suplhur emissions, but

15 P. Crutzen, 'Albedo enhancement by stratospheric sulfur injections: a contribution to resolve a policy dilemma?', *Climatic Change* 77 (2006), p. 211-212. Pdf file available on link.springer.com/article/.

now high up in the stratosphere, so that it will not affect the health of people. It is in a way a natural process. Volcanoes belch large quantities of sulphur compounds. It has been calculated that the reflective power of one kilogram of sulphur in the stratosphere neutralizes the greenhouse effect of 100.000 kilogram of CO_2 emissions. An elegant way to inject the stratosphere with sulphur is available: one could mix some sulphur into the fuel of jet planes.

Before adding contaminants to the atmosphere, one should be certain that side effects would not be worse than the problem one is fighting. In principle it is safer to tinker outside the earth atmosphere. One idea is to install gigantic mirrors or sunscreens in stationary orbits around the earth. Another suggestion is to create a shell of dust particles around the earth, but that would mean the end for earth based astronomy. A more elegant idea is to block sunlight not close to earth but to intercept it at the so-called Lagrange point L1, about 1.5 million kilometres away from earth, four times further than the moon. An object at that point always stays exactly between the sun and earth as it orbits the sun. It would be a major joint effort for the space agencies of Europe, the United States, Russia, Japan, China and India to install a sunscreen at L1, but that would in any case be much more useful than the International Space Station which since many years makes its listless rounds in the outer regions of the earth atmosphere, mainly because world leaders wanted a symbol of international cooperation in space.

Interventions that reduce the amount of sunlight absorbed in the atmosphere can have effect quite rapidly, but they are just stopgaps. Tackling the real problem, the excess of CO_2 in the atmosphere, would be a huge undertaking. It took decades before the warming of the biosphere due to the rise of the CO_2 concentration was sufficiently evident for it to be recognized by almost everybody as an indisputable phenomenon. It would take equally long for reduction of the CO_2 concentration to produce significant cooling.

That slowness of response is caused by the oceans. They have tempered the temperature rise of the atmosphere, but they will also slow down any cooling. Let us first see which effect the oceans have during the process of warming up. There are two mechanisms at work. In the first place, a large part of all CO_2 produced since the Industrial Revolution has been absorbed by the oceans. The concentration of CO_2 in the air would otherwise have been much higher than it is today. In the second place, the sheer presence of the oceans causes a certain increase of the atmospheric CO_2 concentration to lead to less warming of the biosphere than would have been the case without these enormous water reservoirs. The reason is that water has a high 'heat capacity', which means that one needs quite a lot of heat energy to make a litre of water one degree warmer.

If one would embark on withdrawing CO_2 from the atmosphere by some kind of mega-trick, the two mechanisms which first tempered the warming effect of greenhouse gases, will again act as brakes, but now on the attempted cooling. The oceans will start to give back to the atmosphere the CO_2 they once absorbed from it. So, in order to neutralize the greenhouse effect created by mankind, one would not only have to remove all excess greenhouse gases that are presently in the atmosphere, but also the enormous amounts that have since the Industrial Revolution been absorbed by the oceans! The second mechanism – the effect of the large heat capacity of water – would lead to the release of much heat energy into the atmosphere from the cooling oceans, slowing down the atmospheric cooling. So, it would be a Herculean task to enforce cooling by tackling the root cause of warming, the high concentration of greenhouse gases in the atmosphere, instead of simply shielding the earth from sunlight.

Nevertheless, let us assume for the sake of argument that humanity will continue to burn fossil fuels far too long and that one realizes at a late stage that a catastrophe is going to happen unless the accumulation of CO_2 and other greenhouse gases is reversed or at least slowed down drastically. Maybe the realization of nearing doom comes when enormous quantities of methane gas escape from melting permafrost areas, or when the melting of the Greenland ice accelerates. How could one go about removing CO_2 from the atmosphere when panic strikes?

One possibility is to use biomass as fuel for power stations, capture the CO_2 produced and store that away deep underground, isolated from the biosphere. Some tests of such installations are under way. The procedure reduces the amount of CO_2 in the atmosphere, because the vegetation used as biomass fuel has during its growth taken CO_2 from the air. Normally, that gas is released again into the atmosphere when plants rot away or are burnt, but in this case it is captured and removed from the cycle.

Another proposal concerns fertilizing the oceans with iron, nitrogen and phosphorus in order to stimulate the growth of phytoplankton. The growing phytoplankton takes CO_2 from the water, so indirectly from the air. The fraction of the plankton population that sinks into the depth of the ocean before it rots away, carries CO_2 away from the biosphere.

The two methods are in principle identical: the process of photosynthesis is used to withdraw CO_2 from the air. They only differ in the way the gas is prevented from returning into the atmosphere. The *5th Assessment Report* of the IPCC contains estimates of the (lack of) effectiveness of the phytoplankton method. The numbers vary quite a lot, but what it boils down to is that one would have to keep fertilizing the oceans of the world for several centuries if one would wish to neutralize the rise of the CO_2 concentration of the last fifty years. That does not sound like a realistic option, certainly not if one is in an acute panic. The conclusion is that there is no real alternative for serious emission reduction.

Politics and society

I earlier pointed out that it is hardly possible to make credible predictions about anything that primarily depends on social or cultural processes or, even worse, on the whims of human nature. That did not matter much when we discussed world population, globalization and climate change, because there the pressure of circumstances is so strong, that moods and sentiments of groups and individuals can at most temporarily influence the course of events.

In this chapter the situation is different. Here I will address some social and political developments in the 21*st* century in a rather general way, not just in relation to the big themes of the three previous chapters. We will have to accept that human unpredictability and irrationality introduce a large degree of uncertainty into this exercise. More than a sketch of principal directions of events will not be possible, and then only on topics where present trends provide some focus or where convincing analogies can be found with past episodes in history. I will in this chapter often formulate statements more firmly than is really justified, but else the reader would have to struggle through sentences filled with 'possibles' and 'probables'. One may still add such words of doubt according to one's own taste.

Less violence

If one is ready not to be too critical, one may perceive a worldwide trend away from violence and towards more peaceful coexistence. I already drew attention to the extraordinary fact that peace reigns in Europe since seventy years, if one does not count the conflicts after the disintegration of Yugoslavia. Several authors have argued that this is more than a lucky accident. Under the short-term alternations of war and peace they think to see a long-term trend towards a more peaceful and civilized society. In spite of the staining of the last two centuries with butcheries like the Napoleonic Wars, World Wars I and II, the Holocaust and the terror regimes of Stalin and Mao, they do detect in the history of mankind a tendency towards more 'peace on earth'.

It is up to debate whether facts really support that broad thesis, but it is certainly true that acts of crime and violence between citizens have in Europe become much less common. The chance to become a victim of murder is today fifty times smaller than it was in the Middle Ages. On the scale of war and genocide it is difficult to detect a long-term trend, for good or evil. It seems impossible to establish an order of gruesomeness among the genocides committed by the Roman Empire, by Germanic tribes and Huns half a millennium later, by the hordes of Genghis Khan in the Middle Ages and by the Nazis and the Soviet regime in the 20*th* century. And where do the religious wars and persecutions of heretics in 16*th*- and 17*th*-century Europe fit in?

Still, it seems true that through the ages something has changed. The actual brutality and cruelty of violence have not changed, but what did change, is that injustice and violence were in earlier times immutable facts of life, whereas in our time peace and some degree of protection by the law are in ever more parts of the world considered the norm.

Steven Pinker has summed up four causes of increasing peacefulness of human society.[16] The first was the establishment of strong states with the monopoly on the execution of justice and force, which made acts of violence of citizens among themselves far less frequent. Second, trade makes others into partners. Harming them means harming one's own interests. Globalization makes this simple fact ever more important. It has been tried to quantify the importance of globalization as a force for peace. According to that study a 10% increase of direct foreign investments on average decreases the chances of conflict by 3%.[17] The third factor listed by Pinker is the fact that the circle of fellow humans for whom we feel empathy is gradually widening, simply because we more often have contact with people living further away, visit other countries or see them on television. Pinker considers education and intelligence as the fourth pacifying influence: the better someone uses his brain, the less he will be guided by aggressive instincts.

It is unproven that the average citizen has become more intelligent in the last one or two millennia, and it is in any case doubtful if intelligence is a quality that more often leads to peace than to war, but the other points made by Pinker seem to be sound. His second

16 S. Pinker, *The better angels of our nature: Why violence has declined* (Viking Books, 2011).

17 Human Security Report Project, *Human security report2009/2010* (Oxford University Press, 2011).

and third points – increase of economic and empathetic connectedness – were already praised by me in the chapter on globalization. .

Empathy, the we-feeling, was in the time of nomadic hunters restricted to the own clan, later was extended to the inhabitants of one's own village or town, then was widened to the people of a region or even a country, and may for open-minded world citizens nowadays include everybody. Universal empathy has become institutionalized in development aid to Third World countries.

There is no reason why the trend towards peaceful resolution of conflicts would be broken in the 21st century. The pressure of the consequences of climate change will off and on certainly lead to ugly bursts of violence between nations and to civil uprisings, but it is precisely the immensity of the climate problem that will force governments and officials at all levels to work together, which in the long run is the best guarantee against brute violence.

The climate and politics

International talks on climate change at government level will in the near future probably retain the format of mass conferences under the auspices of the United Nations, with heated debates about the wordings of the declaration to be read by that year's chairman, but with few real results. Much more can in fact not be expected, because these conferences are witches' cauldrons in which delegates from all 195 signatories of the Climate Treaty (in full: the United Nations Framework Convention on Climate Change , UNFCC) take part.

The Climate Treaty was drafted in 1992 at the 'Earth Summit' in Rio de Janeiro. It stated the goal of stabilizing the atmospheric concentrations of greenhouse gases in order to 'prevent dangerous anthropogenic interference with Earth's climate system'. The treaty did not include any mechanism to enforce proper measures. In particular, no binding limits on emissions were imposed. The lack of limits and sanctions was precisely the reason why all member states of the United Nations were ready to sign the treaty.

Concrete agreements on emission reduction by developed industrial countries until 2012 (later prolonged until 2020) were laid down in 1997 in the Kyoto Protocol. For developing countries one went no further than to state in general terms that they too ought to aim for emission reduction. Polluter number one, the United States of America, has for that reason refused to ratify the protocol. This behaviour has contributed to the fact that other important polluters have also not been eager to clean up their act.

To obtain stronger results, it will be necessary to establish an organization, preferably under the wings of the United Nations, with a firm mandate that would give it the kind of power enjoyed by the International Monetary Fund or the International Atomic Energy Agency. However, influential big polluters, first of all the USA and China, will for the time being prefer not to commit themselves in any serious way. They will for that reason want to stick to the periodic monster conferences. They will only accept a strong climate organization when in their own country the public mood has changed. That will only happen after the climate change has caused large-scale harm in these countries, in such a way that no reasonable person could deny the causal link. With some luck such disasters will occur already within in a few years, but it may equally well become a wait of one or two decades.

In any case, measures to stop the warming of the biosphere and management of the consequences of the temperature rise will dominate world politics during a large part of the 21st century. Once everybody realizes that emission reduction is an urgent matter, agreements on schedules and sanctions will in principle be reached, but implementation is bound to lead to heated conflicts. Necessary changes in the production and transportation of goods and in the routines of daily life will in some countries be more radical, more expensive or more contrary to existing customs than in others. There will constantly be countries requesting for modification of rules or for special status.

Given that such bickering will last for decades, and given that there will be an even longer aftermath of technical, economic and demographic adjustments to the warmer climate, it is obvious that new alliances will take shape in world politics and some old ones will disappear. Sometimes these may be nothing more than coalitions in a round of talks on a single topic, but new blocs of permanently cooperating nations will almost certainly also arise. It is not possible to make reliable concrete predictions about the birth of new alliances, apart from the truism that surprising political friendships will come and go.

Only the starting position is quite clear. The USA, the nation with by far the largest per capita emission of greenhouse gases and dissident at the time of the Kyoto protocol, will have to make its citizens accept radical changes in their way of life. To win time for that difficult task, this country will at first still try to delay or soften binding agreements. The reputation of the USA will suffer seriously by that behaviour. China, the second largest polluter as a nation (and closing in quickly on the USA) but per capita still a far lighter transgressor than the USA, was in the Kyoto protocol still counted as a developing nation without any specific obligation to reduce emissions. There are clear indications that the Chinese leadership recognizes the seriousness of the climate crisis and that it understands that a calamity cannot be avoided if China behaves as short-sighted as the USA. I would not be surprised if China, Europe and Japan will position themselves as a constructive alliance in climate affairs. Chaotic India – soon to become the most populated country on earth – and unstable, backward Russia may side with the USA. The OPEC countries would be natural allies for this bloc.

What I sketched here, is just an outline of possible positioning in the early phase of serious negotiations. Surprising manoeuvres will follow, but in the end the insight should win that the world is condemned to true cooperation. Some devastating hurricane seasons at the American coasts could be helpful to further that insight.

International negotiations on climate change will profit from the fact that the world has in the past decades become used to international intervention in several areas. Much of that was related to economic globalization, from patents to fishing rights, but also successful activities by institutions like UNESCO, UNICEF and the World Health Organization (WHO) have contributed to a wide acceptance that 'everybody' may put in a word about 'everything'.

The point of view that respect for national sovereignty forbids interference in the affairs of other people, even if the most terrible abuses are known to exist, is becoming ever less popular.

International order

Growth of international cooperation will be forced upon governments because almost all of the truly important problems will be of a global scale, but the establishment of a world government is not to be expected in the 21st century. That would even be undesirable. Differences between socio-cultural, economic and political systems are still so deep, that little good could come from attempts to force that all into the same mould. Accept the differences and negotiate rationally about questions of common concern, that seems the best strategy for the time being.

What will happen is that international organizations dedicated to specific fields of activity become ever more important. They will be granted increasingly stronger mandates when they are called into action. Still, that is quite different from the transfer of sovereignty that countries would have to accept if a world government were instituted. A mandate is no more than empowerment to fulfil a specific task.

While transfer of sovereignty by nations to a 'higher' authority will on a global scale not happen in the 21st century, it already *is* a reality in a number of areas within the European Union, although many politicians do not have the courage to say so to their voters. I will devote a separate section to the EU. Here I only want to recall that not only member states of the EU transfer autonomy in certain areas of legislation. Several other countries conform their rules and regulations to EU standards for economic reasons. Switzerland is an extreme example of that procedure. Economic blocs have also been formed in North and South America and in Asia, but it is unclear whether this will lead to the degree of economic integration that the EU has realized, and it appears almost excluded that they would already in the 21st century follow the European example of abandoning national sovereignty in other areas than the economy.

What will be the hot potatoes in world politics, after the principal issue of climate change? First of all, there will be a host of problems that derive from the climate problem. There will be much conflict material in that area, especially because regulations for the control of consequences of global warming will often collide with the trend towards further globalization. For instance, international agreements for the control of mass migrations due to climate change could interfere with the ideal of worldwide mobility of highly qualified professionals.

Scientific and technological developments will in the 21st century increasingly ask the attention of political leaders. It will often concern dilemmas that in the end boil down to the question of allowing or forbidding the application of a novel technique. The possibility to tinker with the human body and human nature, either by genetic manipulation or by adding artificial body parts, will probably whip up feelings to such a degree that they will call for worldwide regulation. I doubt whether that will indeed result in some kind of global legislation, but the issue will in any case be repeatedly discussed. To the subject itself – improving man beyond his natural capacities – I will devote a separate chapter.

Could there be another world war? Globalization is in general a peace maker, but that cannot be said of global warming. It is a common problem, which only can be solved by uniting forces, but the interests at stake for different nations and world regions vary quite a lot. Some international agreements may by some countries be considered so harmful to their particular interests, that armed conflict breaks out. The risk that one such conflict would grow into a world war nevertheless seems small, thanks to the global interdependence of national economies. War on a grand scale would be too costly for too many parties. This actually means that a world war will not happen again, as long as the world is not ruled by mad people.

Unfortunately, there nevertheless is a large risk that somebody will at some moment use a nuclear weapon. Such weapons are already in the hands of more than one unreliable regime. Further spreading from such murky sources to other suspect states or even terrorist groups is in the long run unavoidable. Therefore, it is only a question of time before a lunatic really throws his bomb. That will not be the end of the world, but it may well be a catastrophe bigger than Hiroshima and Nagasaki combined.

The political reaction will of course be that a firmer line is taken with rogue states like North Korea, but there will not be a total ban of nuclear weapons. The big nuclear powers will want to maintain an arsenal of some size, for the good reason that one should be able to outgun nations or groups of villains secretly developing nuclear weapons. An effort will be made to make inspections watertight, but there will be no guarantee that proliferation can really be stopped. In other words, it may happen more than once that somebody launches a nuclear weapon.

European integration

In the introductory chapter on the sense or nonsense of predictions I called the banking crisis of 2008 and the euro crisis that followed just incidents. In the year 2100 they will be completely forgotten. For today's observer of events, the euro crisis is mainly of interest as proof that European integration has after the fall of the Berlin wall moved on a bit too fast en too rashly. The French hurry towards introduction of the euro as a counterweight against the threat of a big German predominance after reunification, was pure politics without an underpinning in the form of economic and financial structures. These had to be erected later, improvising along the way as the euro crisis unfolded. It is ironic that German pre-eminence could not have been demonstrated more clearly than has been the case in the process of saving the euro.

It is precisely because of these improvised interventions that the net effect of the early introduction of a common European currency can be seen as positive. The problems around the euro forced the members of the euro zone to take irreversible steps towards integration, which is the best route for Europe to take. The disorderly way in which decision were taken, must be counted as a definite minus. In the haste, steps taken were not properly explained to the European citizens, with the result that they felt overwhelmed. That is one reason why the capricious part of the European electorate will in the aftermath of the euro crisis for several years be looking for salvation among populists and extreme nationalists. They will drift back from that dark corner when it has become evident that no alternative worth pursuing is to be found there. Peace and prosperity result from integration, not from isolation.

There will be more shocks on the way to further European integration. Some countries may leave the Union and during some decades the Union may, formally or informally, consist of several layers, the euro zone in the centre, an outer layer of some East European and Balkan countries and the rest in between. These are essentially symptoms of growing pains. When the Union reaches adulthood in the second half of the 21st century, its structure may resemble the Swiss or German model of cantons or federal states with a high degree of autonomy under a federal European government.

Much of what is happening nowadays on a European scale is quite similar to the way in which some present-day nations were formed from smaller precursor states. Switzerland, The Netherlands, Germany, Italy, each of those countries is the product of the merger of a substantial number of formerly sovereign territories into a single nation. As an example I expand a bit on the case of The Netherlands, which I know best. In the 17th and 18th century the country called itself the Republic of the Seven United Netherlands. It was a federation of sovereign provinces, in many ways comparable with the European Union in the beginning of the 21st century. The fact that the provinces really were sovereign is for instance demonstrated by the decision of the province Gelderland in 1787 to grant free passage to Prussian troops on their way to subdue the province of Holland, which had insulted the king's sister. The summed powers of the Dutch federal institutions were quite similar to those of today's European Council, European Parliament and European Commission. The contributions of the provinces to the federal budget (contributions to 'The Hague') were based on their economic weight, in the same way as today's national contributions to 'Brussels'. There was a common currency. Each province issued a certain amount of coins under the supervision of a federal body, entirely analogous to the issuing of euros by national banks under supervision of the European Central Bank. What the Dutch Republic had as extras, compared to the European Union, was the office of stadtholder – a federal, supposedly neutral position always held by a member of the House of Orange-Nassau, sometimes with monarchic tendencies – and a joint army. One tried to conduct a common foreign policy, but that was not always possible. Again taking unruly Gelderland as an example: in 1785 that province did not sign a treaty of friendship with France, while the other provinces did.

Late in the 18th century it became a subject of heated debate if the Dutch provinces should continue as essentially sovereign miniature states within the loose envelop of the Republic of the United Netherlands, or whether it would be better for them to merge into a single nation. Under French influence the road to unification was taken.

In spite of the similarities between the Dutch Republic and her seven provinces on one side and the European Union and her member states on the other side, there is an important difference between the two integration projects. The differences in cultures and languages make relations within the Union less tightly knit than they were within the Dutch republic. Therefore, the end product of the European integration process must be more loosely structured. Dutch unification has resulted in a strongly centralized state, which cannot work on the European scale. Instead, the nations in the political heart of Europe are moving towards something between a confederation and a federal state.

Notwithstanding the differences of context and scale, arguments used by supporters and opponents of unification of the Dutch provinces two hundred years ago, can now be heard on the subject of European integration. In similar situations, instincts and sentiments tend to manifest themselves in similar ways. That is why it helps to have some knowledge of the complexities of national histories, now that we are witnessing the establishment of common institutions between those nations. Of course, it remains true that history does not repeat itself. Even if one could, by some smart algorithm, make corrections for the fact that Dutch unification from beginning to end took place within tighter constraints than European integration, one still could not predict the outcome of the second process on the basis of the outcome of the first.

The Confoederatio Helvetica (the official name of Switzerland) is a more suitable example than the Republic of the Seven United Netherlands if we are looking for an impression of where European integration might finally lead to. In that stable and prosperous state, precisely those circumstances prevail or prevailed which according to sceptics would make a European federal state undesirable and impossible, namely linguistic and cultural diversity – in that respect the cantons Genève and Uri differ not less than Portugal and Latvia – and until recently large differences in wealth between cantons. The number of cantons (26) is about the same as the number of member states of the European Union (28) in 2014. That large number does not prevent effective government. Each canton has its own constitution, parliament and government and has powers of decision on any subject that is not by the national constitution entrusted to the federal government. In short, both European federalists and worshippers of national sovereignty should with some flexibility be able to live with the Swiss model. Therefore, I consider it likely that late in the 21st century the European Union will in its administrative and governmental structure be a large copy of Switzerland.

Forms of government

In the Western world it was until recently thought that our political system of parliamentary democracy operating in a free-market economy is so superior, in principle and in practice, that it would be the best choice for all people in all circumstances. Every nation that could choose freely, would embrace parliamentary democracy and would instinctively know how to handle that difficult system. Who does not remember the surprise of president Bush junior and his entourage when democracy did not spontaneously flourish in Iraq after the fall of Sadam Hussein. And how many did not have similarly unrealistic expectations at the time of the so-called Arab Spring in 2011. After the dictators were chased away, the Arab World would enter an era of political freedom modelled after Western democracies. People who cautioned that the cultural, religious and governmental traditions were much closer to autocracy, were grumps.

An example in a different class is Singapore. This is not a primitive dictatorship or a failed state, on the contrary, it is a very prosperous, well organized city state. However, that orderly state of affairs was reached and is maintained by a strict regime that does not hold civic liberties in high esteem. Criticizing Singapore is in much of Europe considered the proper thing to do.

The economic rise of China and other East Asian states has made their governmental structure into examples for developing nations. The Asian countries have political systems that formally are very diverse, from the monopoly on power of the communist party in China to more or less well-functioning multiparty systems in Japan, South Korea and Taiwan, but they have in common that collective interests have precedence and that they are less concerned about individual freedom and rights than Western democracies. The difference is quite natural, the cultural inheritances of Asia and Europe are just not the same.

We may be disappointed that the ideals of Enlightenment are not (yet) globally shared, but it is a fact that Africa and the Islamic world have more traditional affinity with autocratic rule than with the Western democratic model. It has been the result of the economic, military and political predominance of the USA and its allies that adoption of the model of parliamentary democracy was until recently seen by third world countries as proof that they belonged to the modern world. Often the system was introduced only formally, without much understanding of how the system ought to function. In many of the so-called new democracies it is thought that the winner of an election has the right to claim all power for himself and his clan and may use state institutions to silence opponents. In countries without democratic traditions, an uncomplicated government model of the Singapore type would probably have led to fewer derailments.

In an optimistic mood one could describe the Chinese model as an enlightened collective dictatorship of mostly well-meaning technocrats, recruited from a caste of ideologically indoctrinated bureaucrats. The antithesis of collective interests versus individual welfare, of government versus citizen, is not felt as sharply any more as say ten years ago. As the Chinese become more prosperous and generally better informed, the authorities become more sensitive to the wishes of citizens and gradually allow more liberties in some area. In that sense there is rapprochement to western principles..

But also in the West changes take place that will narrow the gap between the East and the West. The authorities of Western democracies are closing in on their citizens, not because they suddenly mistrust them, but as an unavoidable consequence of electronic storage of the many kinds of information public organizations gather in the course of their work. The ease with which digital databanks can be searched and connected, is an invitation to use data for other purposes than they were collected for. The reasons why authorities do so, are sometimes so commendable that one cannot really speak of abuse: fighting terrorism and crime, hunting for fraud, et cetera.

All this is not really new. Internet users already knew that search engines and providers of social media (and maybe also intelligence services) recorded their movements on the net. It is a step further when we also feel followed and analyzed by public organizations that exist to serve the citizen (instead of spying on him), but it is an unavoidable consequence of the digitization of our society. We will have to learn to live with the fact that institutions will build a virtual shadow of each citizen with ever more information about him. The privacy domain of the citizen will shrink, 'Big brother will indeed be watching'.

Again, that is not absolutely new. After an age in which townspeople could live their lives rather unobserved, they are thrown back to the lack of privacy of life in a village. Their consolation is that one's affairs will not be so entirely out in the open as used to be the case in small communities of the past. All people of the village knew everything about everybody, but in the future only institutions will be omniscient, not people.

In summary, we need not fear the establishment of an evil global dictatorship in the 21st century, but it is also true that the Western model of government, with its emphasis on freedom and rights of the individual citizen, is under some attack. From inside, privacy is under threat, and elsewhere on the globe the East Asian examples of governing a country will gain popularity, at the expense of the Western model. From the perspective of the West that is regrettable, but the reality is that the majority of the world population has more affinity with autocratic government and the simplicity of a single party state than with the nuances of a truly democratic system where minorities have a voice.

Autocracy is the way humanity has organized itself since very early times, starting with the alpha male of the hunting clan and culminating in the absolute monarchies of Europe and Asia. The subtle political game of the multiparty system is even in Western Europe, where it was invented, not yet understood and valued by quite a number of people. Populism is in fact the call for a strong man who shares the instincts of the common people and does not bother with cumbersome democratic procedures.

If at some moment the climate crisis would necessitate fast and forceful measures, that could be an additional reason for the democratic system to become unpopular. A democratic system is by its very structure slower in its reactions than a state where just one person or one clique decides. On the other hand, decisions reached in a democratic process are often better considered. But the merits of a policy decision only become clear afterwards, not in the moments of hurry and excitement. If the warming up of the biosphere should indeed come to be seen as a global urgency requiring immediate, forceful measures, it is to be hoped that the mandate to take these measures will be given to a supranational body whose members are wise experts – not politicians – operating outside the direct control of national governments and parliaments.

Science and technology

Historical perspective

The main message of the previous chapters is that humanity is likely to pass through the next one hundred years without global catastrophes. In the seventies and eighties of the last century, the time of the report of the Club of Rome, the biggest worry about man's future was that the explosive growth of the world population would continue until stopped by famine and disasters due to overpopulation. That nightmare has gone. There will be a soft landing: the number of people on earth will in the second half of the 21*st* century hardly grow anymore and may in fact start to slightly diminish. Climate change has become the biggest problem. Significant warming of the biosphere is inevitable. In the year 2100 the average temperature at the surface of the earth will be at least two or three degrees higher than in the year 2000. The problems that brings, will be large but surmountable. People and societies will after quite some struggles manage to adjust.

I have indicated that one could think of methods to cool down the biosphere to some extent, but it seems unwise to actually try that, mainly because of the danger of unforeseen side effects. The climate is too complex a system to experiment with. Therefore, managing the effects of climate change will not primarily be a matter of futuristic technologies. It will be more important to carefully stimulate natural recovery of the environment.

There are many other areas though where innovative science and technology *will* play a crucial role in the next hundred years. Progress of science is, in contrast to the growth of the world population, really a case of exponential growth. That kind of growth can be maintained for a long time still, because the expansion of knowledge and know-how essentially is a matter of wit and ingenuity, hardly of the availability of natural resources. Science is a self-nurturing phenomenon, like any activity where knowledge and experience is transmitted from one generation to the next. One stands on the shoulders of one's predecessors and so can reach higher.

Never before were so many researchers with so many means active in so many areas. Some of the fields of research are only several decades old and already now producing interesting technologies. Undoubtedly, other directions of research with great application prospects will be started in the coming years, but nobody knows in advance which that will be. In other words, it is impossible to predict the great scientific breakthroughs of the 21st century.

This does not mean that nothing can be said about general *directions* of research in the coming decades and the *kind* of technical applications they might bring about. Sometimes I will even focus on specific results that appear likely to me – the invention of a particular kind of apparatus, the solution to a concrete problem – but the emphasis is then always on the word 'possible', even if I didn't write it down.

In order to get some idea of how fast science and technology will progress in the coming century, it is good first to look back. Only when one has seen how the pace of discovery and invention has accelerated over the past centuries, can one have some inkling of how much will happen in the course of the 21st century.

Technological progress was extremely slow as long as our ancestors lived as nomadic hunters and gatherers. The way one sharpened the head of a spear or a fist axe, started a fire or made pottery could stay unchanged over thousands of years. The transition to a settled way of life – in Mesopotamia already approximately 12,000 years ago, several thousands of years later also around the Mediterranean and in China, the Indus valley, Peru, Mexico – brought social and cultural changes that are often referred to as the first agricultural revolution. The economy of the settlements was based on agriculture and keeping cattle, but the new way of life also led to the rise of trade and specialized craftsmanship. In those times man learned how to use metals, invented writing and money and established the first complex forms of government.

It took a long time before the next big step forward was taken. For thousands of years technical know-how remained at the level of craftsmanship, without any real understanding of the laws of nature. The Greeks and Indians of antiquity were skilled mathematicians, but they did not get far in natural science. Until the end of the Middle Ages there were fundamental misunderstandings about elementary principles of force, velocity and acceleration. As a result, great advances in science and technology were not possible.

The scientific revolution began in the 17th century. That was a Pan-European undertaking. Isaac Newton (GB), Christiaan Huygens (NL), Rene Descartes (F),

Galileo Galilei (I), Gottfried Leibnitz (D) and many more creative minds laid down the foundations of modern science. However, their achievements were not immediately followed by astonishing technical applications. For example, the research of Huygens into the principles of mechanical oscillations led to the pendulum-clock, which because of its precision was of great use to seafarers, but that is rather meagre compared to the potential of other ideas in his notebooks, such as the principle of the steam engine. Most of the tools and routines necessary to exploit the new scientific insights for technical inventions were still lacking.

The large-scale practical exploitation of science began around the middle of the 18*th* century. At that time England saw the start of the Industrial Revolution, the mechanized mass production of goods which in due time would substantially improve the living conditions of broad sections of the population. This technical, economic and social event soon spread to the countries of Western Europe and to North America, and somewhat later to other parts of Europe, but there it stopped. In earlier times China and India had in several respects been more advanced than Europe (everybody knows that gunpowder is a Chinese invention), but the step from craftsmanship to industrial application of technoscience was not taken. There has not been an indigenous industrial revolution in Asia. China's recent spectacular economic development is, through the economic globalization of the past decades, a late fruit of the Industrial Revolution that already in the 18*th* and 19*th* centuries swept through the West.

In the 19th century science became professionalized and institutionalized. In many areas relations of reciprocal promotion arose between science and technology: a technical invention enabled a line of research, which in turn became the basis for a new invention, et cetera. The greatest scientific achievement of the century was achieved by the Scotsman James Clerk Maxwell. With his theory of electromagnetism he succeeded in capturing all electric, magnetic and optical phenomena in a single set of equations. His theory among other things predicted the existence of radio waves. Telecommunication, informatics and all other technologies using electromagnetic phenomena have their origins in the work of Maxwell and the 19th century experimenters who provided him with experimental data. (Maxwell's achievement is far more impressive than that of his contemporary Darwin. It required not much more than close observation and logical thinking to come to the insight of evolution by natural selection, which is why there were others who developed similar ideas as Darwin. Maxwell, however, performed a tour de force at the level of Newton and Einstein.)

The first quarter of the 20th century brought two revolutions in physics. Einstein showed that space and time are elastic projections of a single four-dimensional space-time. This space-time is itself not a rigid medium but is bent and twisted by what is inside it. The other revolutionary new insight was, that the world behaves at its smallest scale – the scale of molecules, atoms and subatomic particles – very much different from what we are used to in our daily life. In the microworld things vary in quantum steps. Something smaller than such a step is not possible. Half steps are not part of reality. Even stranger: the microworld is not deterministic; the outcome of physical processes is at the smallest scale not precisely predictable. We have touched on this subject already when we discussed the sense and nonsense of predictions.

The theory of relativity of Einstein and the laws of quantum physics have meanwhile been verified to extremely great precision. They find application in many ways. GPS navigation functions correctly thanks to the fact that position calculation from the timing of satellite signals takes into account the relativistic difference between passage of time on the earth surface and in the satellites orbiting the earth. Quantum physics is of course used in atomic and nuclear physics, so it is essential for the design of lasers and nuclear reactors, but it also starts to find application elsewhere. Certain microelectronic switching elements make use of the quantum effect that an electron, faced with an energy barrier that according to classical laws of physics is too high to be passed, nevertheless has a chance to 'leak' through.

Key topics in technology

Nobody can know which radically new scientific insights will be born in the 21st century. Revolutionary developments, like quantum physics and the theory of relativity in the 20th century, can by their very nature not be predicted. We can only try to see forward along paths of which we more or less know into what kind of territory they are leading. Fortunately, that is not a very severe handicap for our goal of exploring how applied science and technology will influence the life of the average citizen in the next hundred years. Experience shows that it takes about fifty or hundred years before fundamental research at the frontiers of science finds its way into useful applications in daily life. So, novel technologies that will be important for the people of the 21*st* century are likely to be fruits of research topics that already existed in at least rudimentary form at the time of writing of this book.

Quantum physics, born in the first two decades of the 20*th* century, is an example of the long incubation time between the emergence of an entirely new branch of science and the appearance of technical applications. The technological use of quantum effects has for a long time remained very limited, but in the 21*st* century it may become one of the most interesting areas of applied research, first of all through its role in the broad field of nanotechnology, but also in more direct ways.

Even more important will be molecular biology, especially the topic of research into, and manipulation of, the genetic materials of plants and animals, including human beings. In this field of research the stage of applicability was reached much faster than in quantum physics. That is illustrated by the fact that in 2003 the Human Genome Project – the determination of the series of 'letters' in the human genetic code – could be declared finished, only fifty years after Watson, Crick and Wilkins determined the chemical structure of DNA.

Robots, autonomous machines that can execute complex tasks without instruction by humans, will in the course of the 21st century become much smarter thanks to progress in the field of artificial intelligence. Research on that topic will cross-fertilize with human brain research.

Man himself will be the most important object of application of new technologies. That will not only be in the context of health care and the investigation of man's physiological and psychological functioning, but also with the aim of endowing him with capacities that he does not possess by nature, such as infrared vision or navigation by ultrasound. However, research on humans as biological entities and on artificial improvement of their natural condition will not be a topic of this chapter. The idea of tinkering with ourselves is so far-reaching that it deserves a chapter of its own.

Before I sketch possible developments within the selected areas of quantum effects, molecular biology and autonomous machines, I must explain why I focus on just these three fields of research. There are so many more where enormous progress is to be expected. Space science, a rather grand but established term for almost any activity outside the earth's atmosphere, is not really a field of technoscience in its own right. Topics like earth observation, telecommunication through a network of earth satellites or the exploration of other planets with robot probes make use of, or contribute to, developments in many different technical and scientific disciplines. Climatology is for a similar reason not on my little list, in spite of our preoccupation with climate change.

I also left out fields of research that have ripened to the stage that no major discoveries or new insights seem to be expected any more, although it may very well happen that the greatest discovery of the 21st century comes just from such a corner. Furthermore, I do not discuss research directions which are unlikely to produce practical applications within the next hundred years. This means that the humanities will be entirely left out of the picture. Neither will I speculate about the great discoveries astrophysics and particle physics may bring us within a few decades. These may provide radically deeper insights into the structure of the world, but technical spin-off is not to be expected within a couple of centuries.

This neglect of scientific achievements without foreseeable practical use does not mean that I find the acquisition of knowledge for its own sake unimportant. On the contrary, the only human achievements of lasting value are knowledge and art, without regard for any applicability, but that is not the subject of this book. This book is about the living conditions of our offspring in the near future, about their personal welfare and about the things and services around them. All that will mainly depend on the way man controls himself and his environment, in other words, on technology.

There is one exception, one discovery that would change nothing in man's material environment but would nevertheless have a strong impact on the self-image of humanity, and thus on the way people live. If extraterrestrial life were discovered, the knowledge that we are not alone would affect the spirit of the age more strongly than most technological advances. It is not totally impossible that in the 21st century a life form will be discovered somewhere in our solar system, but I prefer to leave this issue for a later chapter, where I try to look ahead by many centuries.

Below I describe probable developments within the three technological key areas of this chapter that may strongly influence people's lives at some time in the 21st century. I do that in an order which brings us ever closer to man himself as a research object, in order to achieve a smooth transition to the next chapter, about improving man beyond his natural limitations.

Quantum wonders

It is obvious that quantum effects are important in the world of atoms and molecules. That will be discussed in the section on nanotechnology. Surprising is that the oddities of the quantum world can be used to build a supercomputer. The operating principle of such a machine is based on the quantum phenomena of superposition and entanglement. Since nothing similar exists in the macroscopic world, they seem like miracles to us.

Superposition of possible states of existence is a fundamental principle of quantum physics. It means that a particle (molecule, atom, electron, whatever) simultaneously exists in all of its possible states until it interacts with another particle. At that moment it must show itself in one of all possible states. Which states are possible, depends of the history of the particle. The act of 'choosing' among those possibilities is purely a matter of chance. I give an example. Electrons have a property called spin, which one may visualize as rotation around its own axis. Two states of electron spin are possible. The spin rotation of a 'left-handed' electron fits to its propagation direction like a left-turning screw, a 'right-handed' electron behaves like a right-turning screw. An electron born in a process from which both left-handed and right-handed electrons can arise, exists in a superposition of these two states until it interacts with another particle. It then manifests itself as either of the two.

The choice between the two possibilities is purely a matter of probability. If the electron had been born in a process that produces equal amounts of left- and right-handed electrons, then the chance that his particular electron will show itself as left- or right-handed is 50/50. (But by chance it might just happen that ten times in a row such an electron manifest itself as left-handed, just as one may several times in a row roll a six when playing dice.) If the electron were born in a different kind of process where ten times more left-handed electrons are created than right-handed ones, then the chances of manifesting itself as left-handed would be ten to one. So, it would most likely show itself as left-handed, but one could never be sure whether it would not choose the 10% possibility of manifesting itself as a right-handed electron.

The quantum computer derives its power from the principle of superposition. Before we discuss how that works, we must introduce the quantum effect of entanglement, which is the other main ingredient of a quantum computer. Entanglement occurs when two or more particles are created or manipulated in such a way that their common quantum state cannot be characterized by describing each particle separately. They must together be regarded as a single entity, the constituents no longer possess their own, separate identities. To make things even more complicated: the common description of entangled particles must also obey the principle of superposition, which among other things means that every possible exchange of roles between the partner particles must be taken into account.

This very complicated situation comes to a sudden end when one of the entangled particles interacts with the outside world. It must show itself as a definite particle with an identity of its own, which means that the entanglement is broken. Also, it must take on one particular state of being, instead of staying in a superposition of possible states. So it must make a definite 'choice' of spin and of other properties that an elementary particle has. This has the effect that the former partners in the entanglement are also forced to 'choose' unambiguous properties. Such an end to entanglement and superposition when particles must show themselves as real entities to the world, is in the jargon of quantum physics in a suitably dramatic way known as the 'collapse of the wave function'. (In the mathematical formulation of quantum effects the wave function represents the state of being of one or more particles.)

Entanglement can lead to remarkable phenomena. That is best shown with a specific example. If a particle without spin decays into two other particles with spin, their spin rotations must be in opposite directions to keep the net spin of the system equal to zero. The two new particles form an entangled pair. It remains undecided which of them rotates left and which rotates right; each is a superposition of the two possibilities. That changes when we *measure* the spin direction of one of them. That particle must then 'choose' to manifest itself as right-handed or left-handed. The outcome of the measurement is unpredictable (both possible outcomes have the same chance), but if one next measures the spin direction of the second particle, one finds that this particle nevertheless 'knew' how the measurement of the first particle had turned out. The spin direction it shows is always opposite to the spin direction of its partner, so that the net spin is zero, as it should be. One may repeat the experiment ad infinitum, it will always be the case that the second particle shows itself as right-handed if the first had chosen to be left-handed and vice versa. It does not matter how the two measurements are arranged in time and space. It even works if the second measurement is executed so quickly after the first that not even light signals (the fastest messengers possible) could tell the second partner the outcome of the first measurement before it is measured itself.

In short, it is an experimental fact that the second particle always 'feels' without any time delay how its partner shows itself to the measuring instruments, irrespective of how far they are separated from each other. Up to the moment of the breaking of entanglement, spatial distance apparently does not matter. Entanglement is a phenomenon outside the restrictions of space and time.

The quantum computer

The memory of a classical computer consists of bits that represent the values 1 and 0. It is one or the other, nothing in between. Not so in the quantum computer. This machine makes use of superposition. The logical unit of the quantum computer, called a qubit as shorthand for quantum bit, is a superposition of the values 1 and 0. The qubit is a hybrid of both these possibilities. As long as it remains isolated from the outside world, it remains uncertain whether it will be read out as a 1 or a 0. That is not a question of human ignorance but a fundamental property of the qubit as a quantum object. In quantum physics all options stay open as long as possible. In the case of the qubit: until the moment of read out.

A system of two entangled qubits represents the superposition of four possible states, with three entangled qubits that increases to eight states, et cetera. (The general formula is that a quantum computer of n qubits can contain the superposition of 2^n states.) All these states are handled simultaneously by the quantum computer, as elements of the superposition. This is the essential difference with the classical computer, which can only handle one state at a time. To grasp the difference this makes, suppose the two computers are pitted against each other in a television game where one has to choose between hundred doors, behind one of which a heap of gold is hidden. The normal computer must try the doors one by one. It has no chance against the quantum computer, which opens all doors simultaneously.

In order to let it perform a calculation, one must submit the problem to a quantum computer by forcing its qubits into a certain initial state. Next the qubits undergo a programmed series of operations by quantum circuits. Finally all qubits are read out. At this stage each qubit has to manifest itself in one of its two pure states, 1 or 0, so the result of the calculation by a quantum computer containing *n* qubits consists of *n* classical bits of information.

How does one build an actual quantum computer? The principle of quantum computing was first demonstrated in 2001. Since then, there have been experiments with a wide assortment of ideas for the realization of qubits. One might for instance use the two spin directions of electrons to represent the values 1 and 0. Steps in calculations then correspond to the manipulation of spin directions. Other options are polarized beams of light, quantum cavities, quantum vibrations in crystals, and more. In any case, a quantum computer does not at all look like an electronic computer.

The most difficult problem for designers of quantum computers is to make sure that entanglement of qubits can be maintained for the duration of a calculation. Entanglement is essential, but can be broken by the slightest disturbance, even on an atomic scale. Therefore, the objects playing the role of qubits must stay perfectly isolated from the outside world until the moment of read out. The search for robust methods of entanglement is one of the most important reasons why one experiments with so different physical representations of qubits.

The quantum computer is still in an early experimental stage, but the first commercial deliveries have nevertheless already taken place. Google decided in

May 2013 to establish a Quantum Artificial Intelligence Laboratory. Half a year later a quantum computer of 512 qubits was installed there. A machine of apparently almost the same specifications has been bought by Lockheed Martin, the world's largest producer of military equipment.

Research in quantum computing is promoted by the governments of the United States and some other countries, primarily with a view to military applications and analysis of intelligence data. The quantum computer could be very good at searching through immense amounts of data, think of the example of the hundred doors. Our thoughts go automatically to the billions of emails and telephone calls intercepted by American intelligence services every month. The power of the quantum computer as a search engine also explains why Google is interested.

Undoubtedly many more applications will turn up, but the inherently most interesting task for the quantum computer will be the simulation of quantum processes in chemistry and in solid state physics. That will be of immense importance to the field of nanotechnology. It is impossible to do that kind of calculations efficiently with classical computers.

Quantum teleportation

Teleportation, the disappearance of an object from one place and its appearance at another place without moving through the space between, is a popular feature of science fiction. According to our present knowledge of physics this will always remain just a fantasy. However, by use of the phenomenon of entanglement it *is* possible to do something which comes close. It is possible to transfer quantum properties of an object at location A to an object at location B without transmission of any kind of message. This is called quantum teleportation. It is almost the teleportation of science fiction, for if B gets all the properties of A, B becomes indistinguishable from A. The difference is that in Star Trek a single person disappears and appears – 'beam me up Scotty' – whereas in quantum teleportation two objects are involved. No object goes anywhere, only properties are transferred. Still, that happens in a curious way, without exchange of messages and without any time delay, irrespective of the distance between A and B.

In 1997 a research group in Innsbruck demonstrated for the first time that quantum teleportation is possible. Since then it has become a hot research topic. The relevance to quantum computing is obvious. Qubits are units of quantum information, so technically speaking quantum teleportation is the transport of qubits, which is a basic activity inside a quantum computer.

We only need a few words to explain how quantum teleportation works, thanks to the fact that we are already acquainted with entanglement. I again use the example of left and right spinning electrons, and entanglement into pairs with no net spin. We will transfer the spin direction of electron A to electron B. The first step of the procedure is to entangle B with a third electron X. Next, X is also entangled with A. In the course of this last entanglement, X will have to adapt to the state that A is in. If A happens to be spinning left, X has to spin right. But because X has previously already been entangled with B, the right spinning of X forces B to start spinning left. So there we are, the spin direction of A has been transferred to B.

Quantum teleportation has meanwhile been demonstrated with several types of atomic particles. Researchers at the Niels Bohr Institute in Copenhagen have even managed to transfer quantum information between clouds of gas atoms, that is, between atomic ensembles approaching macroscopic size. Although distance does in principle not matter, large spatial separation is at the present state of the art still a problem because of the many sources of disturbance outside the laboratory. Nevertheless, in May 2012 quantum teleportation succeeded between two of the Canary Islands, 143 kilometres apart. This is already an interesting distance when one thinks of applications outside the quantum computer, as it is nearly the minimal height at which satellites can stay in orbit.

The quantum computer and the technique of quantum teleportation are still in very early phases of development, but progress is so fast that they may within a few decades become mature technologies. However, that is not the main reason why I discussed them in some detail. I wanted to show that science has arrived at a stage where quantum physics, the physics of the very smallest, starts to manifest itself very directly in the macroworld of our daily lives. It is very likely that the mastery of quantum physics will in the 21st century produce more inventions of the calibre of the quantum computer.

Nanotechnology

Nanotechnology is about matter on the scale of atoms and molecules. In the words of Nobel prize winner Richard Smalley: 'The grandest dream of nanotechnology is to be able to construct with the atom as building blocks.' In the definition used by the American National Nanotechnology Initiative, it concerns the manipulation of structures that in at least one dimension have a size between 1 and 100 nanometre (nm). There are one million nanometres in one millimetre. That means that one nanometre is to one meter what a marble is to the entire earth. Atoms have diameters of about 1 nm. The upper boundary of 100 nm in the American definition of the nanoworld approximately corresponds to the onset of phenomena that we do not know in the macroscopic world. For instance, below the 100 nm mark aluminium is inflammable and gold, which we know as an inert ('noble') metal, can at the nanoscale act as a powerful chemical catalyst. Nano-objects are not detectable with optical devices, not even with a very strong microscope, because they are smaller than the wave length of visible light, which stretches from 380 nm (violet) to 780 nm (red). Light flows undisturbed over a nano-object, like a wave of water over a small pebble.

Observation and manipulation of individual atoms is possible with the scanning tunnelling microscope. This is not really a microscope, but a surface sensor analogous to the needle of an old-fashioned record player. An extremely fine needle, with a tip that is just one atom wide, slowly scans the surface of the object under study. The needle is kept so close to the scanned surface that the quantum uncertainties of the positions of needle and surface have some overlap. (In quantum physics, nothing has an exact position, there is always some 'haziness'.) The result is that some electrons can jump from the needle to the surface, which generates a tiny but detectable current. The amount of current strongly depends on the precise distance between object and needle. From the current variations one can with atomic resolution reconstruct the hills and valleys of the scanned surface. The same needle can also be used to push atoms.

A more recent invention is the atomic force microscope. It also uses a needle with an extremely thin point, but instead of measuring an electric current one measures movements of the needle. Atomic interactions make the needle slightly wobble as it scans a surface. That wobbly movement is magnified in the reflected image of a laser beam directed at the needle.

Given that by definition nanotechnology includes all research on objects within certain size limits, it goes without saying that it includes very diverse research topics. I will touch on three main branches.

1. Materials with special properties will find applications in micromechanics and microelectronics. A material by the name of graphene is at the moment much in vogue, but there are and will be many others. Graphene consists of single layers of carbon atoms, arranged like microscopic chicken wire. It is one of the strongest materials known. Electrically it is a semi-conductor, like silicon. It may become a base material for fast switching elements in microcircuits.

Applications of this kind of nanotechnology will soon surround us unnoticed, hidden inside pieces of equipment. The sensor that activates airbags in cars and planes is an electromechanical chip containing a nanoscale deceleration detector. It essentially works as follows. Under sudden deceleration of the vehicle the relative positions of minuscule parts of the device change, which in electrical terms means a change of capacitance. The electronics in the chip immediately senses that and issues the order to inflate the airbag.

2. Molecules built from certain chemical elements can under the right conditions spontaneously arrange themselves into specific spatial structures. Such molecular self-assembly is at the base of the formation of macromolecular units in living cells. One can expect that along these same principles nanostructures can be manufactured for applications in electronics, optics, medical applications, et cetera. Whether it will already happen in the 21st century that microchips assemble themselves from molecular components, remains to be seen, but the example of the biological cell shows that something like that should be possible. A microprocessor is less complicated than a living cell. It has probably taken hundreds of millions of years for the first spontaneous assemblies of organic molecules to evolve into a living organism, but that does not mean that we would have to wait terribly long for the first self-assembling chip. The emergence of life forms was a random process, which cannot be compared with the efficiency of smart science.

Before one can think of self-assembling electronic nanocircuits, first of all an equivalent on nanoscale of the transistor must be invented. The transistor is the basic element of almost all electronic circuitry. Its function is to control the amount of electric current that passes, like a valve in a water pipe. Experiments are under way with certain kinds of macromolecules that could perform such a task.

In the world of macromolecules one is almost at the smallest scale where electronic circuits can still function according to the laws of classical physics. The region where quantum leak and other quantum effects start to take over, was entered in 2008 by the research team of Nobel prize winners Andre Geim and Konstantin Novoselov, the discoverers of graphene. They used that material, consisting of just a single layer of carbon atoms, to create a so-called quantum dot – an island of material that due to its tiny size possesses quantum properties – with a diameter of about ten atoms. Such a structure could in principle be used as a transistor. In 2012 the same team realized a functioning transistor based on quantum leak by sandwiching a layer of insulating material between two layers of graphene.

The interest of this kind of experiments is not in the details of how this or that group solved or circumvented this or that technical problem. What matters is that the community of all researchers working on a certain question – in this case: how can electronic circuits be miniaturized down to the scale where quantum effects rule – builds up sufficient width of experience and depth of understanding to finally arrive at a useful solution. Besides the work with graphene, there are so many other promising developments that it seems almost certain that 'quantum chips' will within a few decades exist, if we assume that the delicate problem of connections, inside the chip and between the chip and the outside world, can be solved. (But at some time there will come an end to the miniaturization of electronic circuits. The finish is reached for electronic computers and such devices when chip elements have been scaled down to consist of only small numbers of atoms. Calculating with photons (light particles) instead of electric currents may then be the next step.)

3. Researchers in the field of nanorobotics aim to produce tiny autonomous machines, possibly by using self-assembling macromolecules. Medical applications are obvious. We can think of robots on patrol inside the body, repairing and cleaning blood vessels and organs, detecting and fighting tumours, neutralizing pathogenic organisms and delivering medication to specific places in the body.

Many experiments are in progress in this area. I will mention a few examples. These I selected because they give a good impression of what is going on, not because they would be the most promising ones, for it is still too early to know that. It is still unclear when it will be safely possible and legally allowed to introduce nanorobots into the human body. Our body is such a complex environment that it would be very risky to launch little robots into the bloodstream without first thoroughly testing their behaviour in surroundings that are easier to survey. That means that one or two decades of testing will be necessary between the present first experiments with nanorobots and their more or less routine use for medical purposes.

In several laboratories nanotransporters are being developed. The idea is that they would navigate the bloodstream to deliver medication to specific addresses, or maybe to transport DNA segments to an organ in need of gene therapy. The transporter has the shape of a capsule, consisting of a number of macromolecules that have joined up to form a hollow ball, inside which the payload is transported. The outer side of the capsule may be coated with 'sleuth molecules' that can detect whether the capsule has arrived near the kind of body cells where its contents must be discharged. Such a macromolecular capsule is the smarter nephew of an already existing device for the deposition of drugs: at present one uses liposomes (tiny, artificially prepared fatty bubbles) to carry pharmaceutical substances to the sites of certain forms of cancer or certain infections.

Another idea is to attack cancer cells with a nanosize cruise missile: a deadly weapon that finds the route to its target all by itself. Size and shape of the weapon could be made such that it would not be able to invade blood cells and other healthy cells it meets on its way, but it would be able to slip through the pores of the membrane of a cancer cell. An option is to let the weapon attach itself to a lymphocyte of the immune system that by its nature is on the lookout for certain kinds of cancer cells. Once arrived at the tumour and intruded, the nano-weapon could blow itself up, or it could carry a magnetic molecule, which could by an external magnetic field be brought to vibrate so violently that the walls of the cancer cell burst.

Photo activation will be a way to issue commands to nanomachines. This technique amounts to including light sensitive proteins in the nanostructure in such a way that certain sequences of laser pulses will start a desired action. Besides nanorobots that let themselves be carried as little ships or capsules by streams of liquids or gases, there also are tiny carts moving on wheels or balls, but they cannot yet move on their own. An engine at nanoscale is still missing.

Still, the present lack of an all-purpose engine doesn't mean that autonomous movement of nanomachines is always impossible. Special environments can offer special solutions. In 2014 researchers of the University of California at San Diego introduced extremely simple nanorockets into the stomachs of mice. These little devices were tiny tubes, open at one end and closed at the other end, with a coating of zinc on the inner wall. The chemical reaction of zinc with stomach acid produced hydrogen gas, which escaped at the open ends of the tubes. The nanoscale jet drive!

Robotics and artificial intelligence

A robot can execute activities of some degree of complexity with a large degree of autonomy, even when conditions change. Being up to some variation in tasks and working environment is what distinguishes a robot from other machines. It is a question of taste where one places the exact border line..

In the last decades much progress has been made in the areas of sensors, manipulators and mobility, and much more is certainly still to come, but these are not the attributes that ultimately determine how well a robot can manage complex tasks. Above all, the performance of an advanced robot depends on its ability to analyze situations and decide on the right response. Machines that do about as well as an insect in those respects, will soon be on the market in a large variety of shapes and sizes.

The driverless robot cars that Google has on the road seem to be functioning safely and reliably. In August 2012 Google announced that their cars, of which there are at any moment at least ten on the road, had together driven more than 500,000 km without a single traffic accident. (A year before, one of their cars *had* been involved in an accident, but that didn't count, because it was at that moment driven by a person.) The Google car sees its surroundings and knows where it is by use of radar, lidar (radar with laser beams), GPS-navigation and digital cameras. It is able to interpret the data it collects 'like a human', in the sense that it recognizes other traffic participants, curb stones, traffic signs and obstacles, continuously evaluates the traffic situation and on that basis chooses its path from A to B. Some professional car manufacturers also have more or less functioning driverless prototypes. Google has the intention to commercialize its technology before the year 2020. Audi, BMW, General Motors and Nissan have announced that they will not be far behind. Some states of the USA already allow robot cars on the road.

It doesn't matter much when exactly the first driverless cars will become available to the general public. The important thing is that a race is going on between developers of a robotized product for a potentially very large market. This means a tremendous boost for R&D in robotics, with spin-offs to many other areas than road transport.

Military applications are another large growth market for mobile robots. Warfare without direct participation of humans in the fighting will soon be a reality in the air, where it is technically easiest and where it quickly saves a lot of money. A high-tech drone is nearly a hundred times cheaper than the latest generation of manned jet fighters. Robotization of ground armies will of course profit from the civil development of robot cars, but large sums of defence budgets will in addition be spent on research into robot mobility in extreme conditions.

For military commanders and for officials responsible for defence budgets automated warfare is an attractive perspective, but not for humanity. The possibility to attack without risking the lives of one's soldiers very much lowers the threshold to armed conflict and increases the danger for civilian populations. Unfortunately, the trend towards robotized attack weapons appears unstoppable. Defensive countermeasures are difficult, because they will have little deterrent effect. The attacker only has machines to lose. Military strategists may conclude that the threat of massive counterattacks is the best defence. That could ultimately mean a return to the strategy of 'mutually assured destruction' (with the apt acronym MAD) of the Cold War between the United States and the Soviet Union.

In short, progress in robotics will bring many benefits and comforts into our daily lives and it will expand the limits of exploration in the oceans, under the surface of the earth and in space, but it also forms a severe threat. Waging war becomes too easy and too cheap. The risky strategy of MAD may be the only way out. During the Cold War of 1945-1989 that worked, but there is no guarantee that a mad dictator wouldn't this time start an apocalyptic series of attack and automatic counterattack. I will return to this question in the last chapter.

Robots that look like people, the 'humanoids' of science fiction who imitate us in appearance and behaviour, are not of interest when we try to explore the future. Such machines may well become available, furnished with pleasant facial expressions and beautiful voices, but they will be just toys or social companions. True robots should *not* look like persons, for they must do things for us that we cannot or do not want to do ourselves. A robot must be designed for optimal performance at tasks that are not suitable for human beings.

The ideal robot, employable in many situations, will have sensors to explore its surroundings, possess several sorts of manipulators to execute tasks and move easily through the world, but above all it will have something akin to intelligence.

People working in the field of artificial intelligence usually speak of AI. Since it is so nicely short, I will adopt their jargon. Robotics is just one of several research areas where progress in AI is being followed with great interest. For instance, brain researchers are interested in similarities or analogies between biological and artificial intelligence. But the attempt to create an intelligent system outside a biological entity is in itself already of such fundamental importance that it does not need justification in the form of applications in other fields.

Since the appearance of the first electronic computers, AI is the big promise that has not yet been fulfilled. It is in that sense in the same league as nuclear fusion – also since long a promise for the near future – but in the case of AI the problem is more fundamental. The principles of nuclear fusion are crystal clear, it is just that the technical realization turns out to be extremely difficult. The problem with AI already begins with the second part of the name: what *is* intelligence? Achievements which once were considered as proof of intelligence, such as defeating the world's best chess player, turned out to be possible by decent programming, brute computing force and a huge data base. That is how Deep Blue of IBM defeated the human chess wonder Garri Kasparov in 1997. Deep Blue did not posses any real intelligence, it didn't know what it was doing. Achievements that only seem to demonstrate intelligence are classified as 'weak AI'. Here I am only interested in strong AI: a machine that can do more than apply the rules of logic and arithmetic.

But what exactly is that 'more'? The impasse has been jokingly but aptly expressed in the definition that artificial intelligence is that which we cannot yet let a computer do. In this state of affairs, researchers have the tendency to concentrate their work on a single, specific quality an intelligent machine should in any case possess. How intelligence would arise from joining many of such qualities, stays unanswered.

I suspect that strong AI cannot be achieved without an ingredient similar to the human qualities of curiosity and eagerness to learn. That might even be both the foundation and the cement of the structure we call intelligence. In this metaphor, the skills that present-day AI research concentrates on are loose building blocks.

Curiosity and other drives and emotions felt by human beings are connected with chemical stimuli the brain receives. In other words, the whole body contributes to human intelligence, not just his brain. (The brain is itself also involved in its chemical steering: a particularly important producer of hormones is the part of the brain known as the hypothalamus.) Research on AI has until now entirely concentrated on simulating the brain as an information processor, an organ that can think logically, and uses the hardware and software of electronic computers for its experiments. The question then is which phenomenon, process or thing could have the same effect on a computer as our body, in its role as a producer of hormones, has on our brain. How to provide a computer with the emotional foundation for intelligence? I wouldn't know. Is the electronic computer maybe too sterile an apparatus to become a carrier of artificial intelligence? Maybe we must change over to hardware with biological switching elements, more biochemistry and less semiconductor electronics.

It may well be that my impression as an outsider is too pessimistic. There are in any case many intelligent, enthusiastic people working in the field of AI and the belief that the earth will one day be ruled by super intelligent computers or robots is very much alive in some circles. Among some there is an almost religious belief that this takeover will already happen around the year 2045.[18] I give them little chance, and I don't just mean the timetable. It seems to me that attempts to develop a superior artificial intelligence, in silicon chips or in any other substrate, will be no match for a programme of improvement of human beings, by genetic engineering or otherwise. We will come back to this later.

Genetic engineering

The technique of purposeful interference with the genetic material of plants and animals will in the coming centuries have at least as much impact on society as the sum of all technologies dealing with dead matter. Many applications already exist, but that is nothing compared to what is in store.

The first case of genetic modification dates back to 1973. That concerned a kind of bacteria. Opponents of genetic engineering often prefer to forget that modified bacteria are since decades being used for the production of medicines – bacteria producing insulin were already used on an industrial scale in 1982 – and for the production of enzymes for washing powders. Viruses have been especially engineered for safe vaccinations: good immunity without a risk of infection. Animals have by genetic manipulation been made susceptible to human diseases in order to be able to test possible cures in a systematic way. Genetically

18 The year 2045 was in 2005 mentioned by Ray Kurzweil in his messianic book *The singularity is near*. While I am writing these words, it is ten years later and artificial intelligence has not come much closer, but Kurzweil has not yet announced that the big event would be delayed to a later time.

modified crops are on the market since 1994. Some were made resistant against insect plagues or virus infections, others were modified for higher nutritious value or for the ability to thrive in a difficult climate, all of which has substantially contributed to the agricultural revolution.

Ever more micro-organisms, plants and animals will in ever more ways be usefully improved, faster and more radically than has been possible in the past with the primitive method of selective breeding. Changing bacteria in such a way that they will fulfil tasks outside of their normal behaviour can lead to spectacular innovations. Some examples already exist, like the cleanup of oil spills and the processing of poisonous waste by bacteria, but much more is possible in that area.

In spite of the benefits genetic modification has already brought, the technique is in Europe still widely rejected by the general public. That is partly nothing more than the well-known phenomenon of fear of new things that one doesn't understand, as once was the case with the train and the automobile, but in the case of genetic engineering there is more afoot. There is opposition out of principle by people who regard it as sinful tinkering with God's creation, and by environmental activists who blindly fight anything that is not 'natural' (but who have a dog that does not look like a wolf at all).

Entirely in line with that negative image, genetically modified products are at this moment (2014) the subject of very restrictive regulations. A European directive from 2004 says that the cultivation of genetically modified organisms is only allowed for research, not for commercial purposes. If a food item contains more than 0.9% of modified ingredients, or if it is not possible to determine whether there are any modified ingredients in it, then a special permit is needed before it is allowed to appear on the market. The procedure for obtaining that permit has been made so difficult and tedious, that very little genetically modified food is for sale in countries of the Europe Union, which is precisely what the legislators wanted. In the USA, China, Brazil and some other countries the cultivation and consumption of modified crops is already wide-spread.

Year after year, evidence is accumulating of how useful and profitable the genetic improvement of organisms can be, and how farfetched horror stories are about the dangers of cross pollination between escaped Frankenstein plants and the plants of mother nature. One day public opinion – and thus legislation – in Europe will under the weight of facts turn around, but I fear that it may take several more decades before that happens. The harm Europe is inflicting on itself must first become truly painful before the sentiment will change.

The genetically modified products that presently are on the market, are strictly speaking not unnatural. The changes made consist of deactivation of certain genes, activation of some genes that in the course of evolution had become turned off, or implantation of some genes from one kind of organism to another. In short, only naturally occurring genes are involved.

That will be different in the emerging field of synthetic biology. Therefore, workers in this area can expect even harsher opposition from green and religiously inspired groups. The truth is that synthetic biology can become of great value, precisely *because* it concerns itself with possible variations on the principles of genetic coding that do not occur in nature. It may result in very useful applications outside the range of what is possible with 'natural' biology. How handy and useful 'unnatural' things can be, has been shown many times, starting with the invention of the wheel.

A very short course on the essentials of genetic coding may help to appreciate the difference between traditional and synthetic biology. Each cell of a biological organism contains a set of DNA (**d**eoxyribo**n**ucleic **a**cid) molecules. These macromolecules carry the genetic material, the genome, of the organism. How is genetic information stored in a DNA molecule? DNA is a very long molecule in the shape of a helix, rather like a winding ladder. It is an assembly of many smaller molecules. Most important are the molecules which form the rungs of the ladder. Each rung is built from two molecules, one attached to one leg of the ladder, its partner to the other leg. In the middle they are loosely attached to each other. In all living creatures the same four kinds of molecule (adenine, thymine, guanine and cytosine) are used to construct the rungs. One usually just writes them as A, T, G and C and calls them the letters of the genetic code. These molecules belong to the chemical class of the bases. The rungs of the DNA ladder are exclusively made of the pairs A-T and G-C. Other combinations, such as A with G or A with C, just do not occur. So, if one knows which letter is on one side of a rung (for instance a G), then one automatically knows the letter on the other side (in this case, a C). The two sides are like the print and the negative of an old-fashioned photo.

A DNA molecule is functionally divided into sections, each section consisting of many consecutive rungs of the ladder. Each section is called a gene. The sequence of letters in a gene is the recipe for the preparation of a specific protein. This is what DNA is: a cookbook with recipes for proteins. The human genome consists of approximately 22,000 genes. Within the biological cell, proteins are involved in everything. They regulate the metabolism and other cell processes, transmit messages, transport materials and serve themselves as construction materials.

How does the recipe for a protein look like? A protein is a molecule consisting of a sequence of smaller molecules, the so-called amino acids. Biology uses twenty different amino acids. In a gene is written which sequence of amino acids must be glued together to make a certain protein. That message is coded as follows. Three successive rungs of a DNA ladder together represent what is called a codon, the genetic code for one of the twenty amino acids. (That can easily be done. With combinations of the four letters A, T, G and C one can form 64 different codons, much more than the twenty used by earthly life forms.) The genetic code is universal: a given codon will in all organisms, from the simplest algae to humans, lead to the production of the same amino acid. This is why transplantation of a gene from one species to another is possible with preservation of the function of the gene.

How does a cell read a recipe from a gene and how does it then assemble the desired protein? The first step is that, over the full length of the gene, the partners of the base pairs A-T and G-C forming the rungs of the DNA ladder let go of each other. The result is that this section of the DNA molecule opens up like a zipper. Next, an RNA molecule is built up along the open zipper, in the way of making a print from a photo negative. RNA (ribonucleic acid) is similar to DNA. It acts as carrier of the protein recipe to a ribosome in the cell body. Ribosomes are very complex examples of natural nanotechnology. They are programmable assembly facilities, themselves consisting of a complicated array of self-organizing proteins and RNA chains. The ribosome reads step-by-step, always three letters at a time, the recipe brought to it by the RNA molecule that acted as messenger, assembles the corresponding amino acids on the spot from raw materials drifting in the cell body, immediately glues the amino acids together and out comes the ordered protein.

Synthetic biology aims to step outside the limitations of natural biology in several ways. That could, for instance, enable the production of 'unnatural' proteins by bacteria or other micro-organisms. The range of applications in medicine and in the production of plastics and fuels could be large. Some start-up companies in the USA are already experimenting with the production of bio fuels by synthetically engineered organisms. They have for example let yeast cells produce pure diesel oil.

The least rigorous intervention is to implant a synthetic gene in which codons have been arranged in such a way that the corresponding sequence of amino acids leads to a protein with interesting properties. Next, one can move one step further away from nature by inducing bacteria to translate a certain codon into an unnatural amino acid. This could be done without harm to the natural protein production of that kind of bacteria, for we already saw that there exist more different codons than are needed for the twenty naturally occurring amino acids. The result is that there are some amino acids that can in a genetic recipe be referred to by two or even three different codons. It has been demonstrated that bacteria can be 'taught' to associate one of the superfluous codons with a new, unnatural amino acid. That lesson was taught by the old technique of selective breeding[19] – which can work pretty fast when applied to fast mutating bacteria – but it may soon be possible to control the translation of codon to protein directly.

The ultimate step towards synthetic genetics would be to add one or more letters to the natural letters A, T, G, and C, or even to switch to an entirely new genetic alphabet. However, it is as yet not clear why that would be of much interest. One would first have to understand not only genetics but the entire molecular biochemistry of the cell, else one would not be able to tune that biochemical factory usefully to a completely new genetic cookbook. We still have a long road to travel before we reach that stage. Until then, it does not make very much sense to go to extremes on the side of genetics.

An idea with more obvious benefit is to construct the legs of the DNA ladder from other material. In

19 A research group led by George Church performed this experiment on the bacteria Escherichia coli (*Science*, October 18, 2013).

nature, the legs are made of sugar molecules coupled to a phosphate. Several research groups have succeeded in replacing the sugars with other molecules. That has no effect on the genetic information, for that is located in the rungs of the ladder. Modification of the legs of the ladder is useful because it means that one can make synthetic organisms prisoners of the laboratory. By choosing sufficiently exotic ingredients for its DNA ladders, the organisms will in the real world not find the food they need to multiply.

Researchers at the J. Craig Venter Institute gave in 2010 their own interpretation to the words synthetic biology. They did not try to make bacteria do unnatural things. Their goal was to demonstrate that genetic engineering is sufficiently developed to synthesize a complete genome of one million bases from scratch – that is, starting from basic biochemical ingredients – and to successfully implant that product into a cell and to make that cell multiply. They built a replica of the genome of an existing kind of bacteria, to which they added as playful notes the coded versions of their own names and an internet address.

Synthetic biology attracts much attention, but even more exciting results could conceivably come from another direction of research in molecular biology. It will be very interesting to see how synthetic biologists can create organisms with remarkable properties by extending the genetic code beyond its natural boundaries, and it is very likely that their work will bring many useful applications, but it is less clear whether it will lead to further scientific breakthroughs. Much of this research essentially boils down to replacing molecules of the sort A by molecules of the sort B, within an already quite well understood framework. How the genetic code is stored in the DNA molecule, how genes are read and decoded, all this is known and adding a new letter to the genetic code is not likely to lead to fundamentally deeper insights.

Truly surprising discoveries might come from a line of research where uncertainty and confusion still reign. Epigenetics is such a topic. This word has in the course of the 20th century had a number of quite different meanings, but presently it stands for research into changes in gene activity due to influences from *outside* the genome. Epigenetics investigates how the chemical environment of the DNA molecule can cause certain genes to become inactive (stop having their code copied onto an RNA molecule) and others to become extremely active (having their code very frequently copied onto RNA), with the result that the organism loses some properties and acquires others.

This means that the question of cell differentiation within an organism is a key issue for epigenetics. When an inseminated egg cell begins to multiply, until the stage of eight cells only perfect stem cells are produced, each of them in principle still able to become a complete organism of its own. In the next generations of cell division differentiation starts. In the mature body almost every cell is firmly programmed to function in a very special way, although all these specialized cells – heart cells, liver cells, nerve cells et cetera – are entirely identical in their DNA. It is a bit like identical triplets, one growing up to become a musician, the second a plumber and the third an accountant. The comparison does not quite hold though, because the human triplets could have chosen other professions, while cell differentiation in the embryo is a fixed process. These has to be a heart, a lever, et cetera. In other words, the *epigenetic* conditions that cause cell differentiation in a growing embryo are themselves governed by instructions that are *genetically* laid down (fixed in the DNA of the species).

Cell differentiation within an organism – different expression of identical genetic properties – is sometimes caused by a chemical influence on a single 'letter' of the genome, in other cases by the way in which the long DNA molecules are packed up inside the nucleus of the cell. The DNA molecules do not just lie around in the nucleus. They are chemically loosely bound to so-called histones. Histones are clumps of proteins that serve as spools around which sections of a DNA molecule are wound – 147 letter pairs per histone. The ends of histone proteins dangle out of the clump and are targeted by enzymes that can make the clump tighter or more loosely structured. In the latter case RNA messenger molecules can easily move in and out, with the result that a gene which happens to lie at that histone is frequently copied. A DNA segment which finds itself on an impenetrably tight histone, will be doomed to inactivity.

The epigenetic changes taking place during the development of an embryo follow a recipe typical of the species. That recipe must in some way or the other be laid down in the genome. However, in the last decade several examples have been reported of epigenetic changes ('epimutations') as a result of environmental influences. What must one think of that?

The epigenetic configuration of a cell stays intact when it divides into two daughter cells. So, in the absence of external disturbances, an individual will during his entire life keep the epigenetic make-up formed in the course of his development as an embryo. However, transfer of epigenetic information from generation to generation would seem to be excluded, because sperm and egg cells in principle have no epigenetic filter around the genome at all. That is why cells taken from an embryo in its very earliest stage of development are ideal stem cells. Their full genome lays equally accessible, no portions of their DNA have yet been deactivated or stimulated.

Until a few years ago, this seemed undisputable. However, recently reports have been published of *heritable* epigenetic traits, in bacteria, in plants and in animals, including man. If correct, those observations would be spectacular, because it implies that experiences of parents can in some cases be biologically transferred to their offspring.

Whatever one may think of these reports, it is unlikely that inheritable epigenetic mutations, if occurring at all, could remain stable in the long run, instead of disappearing after a few generations. The reason is that the biochemistry of epigenetic change consists of reversible processes. This 'soft', temporary character distinguishes epigenetic mutation from the 'hard', durable genetic mutations. If epigenetic mutations were in some species really lasting, the biological evolution of those species would have to be very fast, since epigenetic change can proceed at a much faster pace than the Darwinian mechanism of natural selection operating on accidental genetic mutations.

The idea that an organism could biologically pass on to its children characteristics it had acquired during its lifetime, had early in the 19th century already been formulated by the French biologist Jean-Baptiste Lamarck. His theory was at one time widely accepted, but was discredited by Darwin's work. Now, almost two hundred years later, Lamarckism might be heading for some degree of rehabilitation.

Experimental evidence for epigenetic inheritance in the case of humans has been reported from The Netherlands. It is the result of an analysis of medical records of children born in an Amsterdam hospital during the famine of the winter of 1944/1945.[20] It is not surprising that the children of mothers who starved during pregnancy were on average smaller than normal children and generally less healthy, but the astonishing find was that the children of these children also were on average shorter than normal at birth (with a calculated chance of only 1:100 that the difference was a pure coincidence) and that in their later lives they suffered illnesses twice as often as others of their age.

It is still unclear how the transfer of epigenetic information from generation to generation could work.[21] This puzzle will probably be solved within a few decades. Assuming that the existence of inheritable epimutation will be firmly established, experiments with plants and animals under controlled conditions will at some time clarify which mechanisms are at work.

20 R.C. Painter et al., 'Transgenerational effects of prenatal exposure to the Dutch famine on neonatal adiposity and health in later life', *British journal of obstetrics and gynaecology* 115 (2008), p. 1243-1249. Online in PubMed under nr. 18715409.

21 V.L. Chandler, 'Paramutation's properties and puzzles', *Science* 330 (October 2010) , p. 628-629.

Because of its reversibility it could become an attractive alternative for the more rigorous intervention of genetic modification.

Tinkering with ourselves

A longer life

During the 21st century remedies against many illnesses will undoubtedly be found, but it is by no means certain whether progress, measured by the number of diseases conquered or the number of human lives saved, will be larger than in the 20th century. At the frontier of medicine no more 'easy' diseases are waiting. The fight is concentrating primarily on cancer in its many forms, disorders of the nervous system (MS, ALS, Alzheimer and Parkinson) and an array of tropical diseases.

Several types of cancer can already be treated. In view of the enormous progress in the field of molecular biology, we have reason to hope that this will in twenty or thirty years be true of almost all sorts of cancer. Cancer is the result of defects at certain critical locations in the genome. Such defects may be due to an inherited mutation, a viral infection or an environmental hazard, such as contact with certain chemicals or exposure to a high dose of ultraviolet radiation. If in future the DNA sequence of every newborn child will be recorded, before or shortly after birth, it will be possible to prevent inherited forms of cancer to develop, or at least to treat them at a very early stage, possibly with gene therapy. I am not able to judge how likely it is that one hundred years from now all forms of cancer can be cured (instead of just kept under control).

Certain aspects of cancer research may turn out to be relevant also for research on aging, or rather, on how to slow down aging. In a culture of normal human cells, division of cells stops after about fifty generations. Thereafter a process of aging begins inside the cells: they structurally decay, their chemistry begins to malfunction and soon halts entirely. We may suspect that the limited life span of multicellular organisms is due to the fact that there seems to be a limit to the number of successive cell divisions. The next question then is what causes that limit and how it could be circumvented. This is the connection with cancer research, for it is the nature of cancer cells that they just never want to stop dividing.

A popular theory holds that normal cells at some moment stop dividing because the two ends of each DNA molecule, the so-called telomeres, become shorter at every division. The shrinking of the telomeres is a fact. The enzymes which take care of DNA duplication during cell division seem not to be able to do their job till the very ends of the molecule, with the result that the telomeres get shorter at every division. When they become shortened to a certain minimal length, division stops. Telomeres contain nothing more than repeating base pairs without genetic meaning, so it does no harm when they become shorter with each cell division. They are disposables, apparently with no other function than to protect the rest of the DNA molecule. The theory says that division stops when telomeres have become so short that in the next division a genetically meaningful piece of DNA would be lost.

Cancer cells have found a way around. The telomeres of cancer cells do not become shorter after repeated divisions, because these cells activate certain enzymes which can make new pieces of telomere grow. Stem cells also possess this enzyme. In experiments with laboratory mice some signs of aging could indeed be reversed when growth of the telomeres was induced, but there is as yet no convincing evidence that the key to a long life lies in the manipulation of telomeres. There are more theories about the cause of physical aging, but it is unclear if any of them is on the right track. Nevertheless, the understanding of cell processes is increasing so fast that it seems very likely that during the 21st century the recipe for a longer life in reasonably good health will be discovered. To put this into some numbers: my guess is that in the year 2100 a ninety year old person will physically and mentally be in the same condition as somebody of sixty today.

Use and abuse of DNA profiling

The application of techniques of genetic engineering to humans is controversial, and rightly so, but not many people would object to just *testing* whether a newborn child suffers from an inherited illness or is genetically predisposed for certain disorders. Such screening programmes will become more and more valuable, since we may expect rapid growth in the knowledge about diseases caused by genetic defects.

It will probably soon become routine in many countries to perform the genetic screening already on the embryo. If it is found that something is very wrong, there are in principle three possible ways to react. One may in many cases decide to start treatment at some moment *after* birth, which might include the use of gene therapy. This means that one would try to repair the genetic error, most likely by injecting a healthy version of the affected gene. There will, however, be cases where it appears medically necessary to already intervene when the child is still in the womb. Refinement of technologies of micromanipulation will make that possible ever more often. The third possible outcome is abortion. That could be the case if the genetic findings and the results of further tests on the embryo show that a severely handicapped, hardly viable child would be born.

In 2012 researchers at Stanford University in California found a method to determine almost the full genome of a foetus from blood samples of the mother. It is, therefore, to be expected that it will soon become standard procedure to record the entire genome of each child before birth, instead of only screening for genetic diseases. It will be a logical step to use this information as the basis of the medical dossier of each new-born child. Knowledge of a person's full genome will indeed become ever more valuable from a medical point of view, given that one will in the next decades learn much about the effects of mutations all over the genome, whereas present genetic screening only concentrates on a small number of genes that are especially known to be related to inheritable diseases. Quite generally, knowing somebody's full genetic make-up will make it possible to fine-tune medication and preventive care. Physicians could prescribe lifestyle adaptations or preventive medication if mutations or defects in certain genes indicate that a person is susceptible to a certain illness.

All of the above can without much reservation be called useful and beneficial. Doubts start when we leave the field of medical applications and think of what else could be done with genetic information. If at some time optional genetic screening for medical reasons would be replaced by an obligation that every citizen should have his full DNA profile recorded, then the possibility of misuse of that information lurks in many corners.

To start on the positive side, an application that might do more good than harm would be for governments to make a person's DNA profile his only legally valid means of identification. Is the encoded profile stored in a chip on somebody's passport, drivers license, credit card and the like, then identification fraud will be almost impossible as soon as handy, cheap machines for DNA sequencing become widely available. The identity check will consist of comparing the DNA profile in the chip with the result of a short touch of the machine at the inside of the person's cheek.

For such a procedure of personal identification it would not even be necessary to use complete DNA profiles. It is good enough to check only some DNA regions of which it is known that they can differ strongly from individual to individual. Nowadays it is customary to combine eleven of these special regions into a person's 'genetic finger print'. The chance it would not be unique, that there would be another earthling with exactly the same genetic finger print, is one against many billions.

In order to identify somebody by her DNA, is it sufficient to know the succession of base pairs ('letters') in the genome. It is not necessary to know what that sequence *means*, how it will express itself in characteristics of that person. Severe misuse of DNA profiles becomes possible when genetic science progresses to the stage where the human genome can be reliably *interpreted*, in the sense that qualities and deficiencies of a person can be deduced from the sequence of 'letters' in her genome. At present, that is possible on a very limited scale only. It does not go much further than the screening for a few inheritable diseases that we just discussed. But once it is known, throughout the genome, which combination of genes programmes a person for which characteristics, it will be possible to deduce from the DNA profile much about a person's bodily constitution, her character, her talents and much more. We can imagine many applications of such knowledge, not all of a desirable nature. It is a dilemma which we have encountered before: strong technologies can be used in many ways, some good and some evil, and not everybody will agree about where the border line is between the two.

For example, insurance companies would like to base premiums on the risk profile exhibited by a customer's DNA profile. In the case of health insurance that would in many countries be forbidden, because it is an insurance founded on the idea of solidarity – the healthy supporting the sick – but many car insurances already work on the principle that people with a poor safety record pay more. Insurance companies would no doubt welcome the additional possibility of premium differentiation on the basis of genetic characteristics. If somebody's DNA profile shows that she is inclined to an aggressive, reckless style of driving, she will find insurance only at a very high cost.

This simple example introduces us to the difficult question of whether a person who misbehaves must pay for his misdeeds, or whether he should be considered a victim of his bad genes and should be treated as a patient. As more becomes known about the way in which genetic make-up determines somebody's actions, we will more often tend to see antisocial or even criminal behaviour as the acts of a pitiful person who had the bad luck to be born with the wrong genes. But a bad start at birth should not be a license to systematic misbehaviour. When at some time in the future the connection between genetic defaults and antisocial habits becomes clear, that same knowledge would probably show the way to therapies. Somebody diagnosed as needing such therapy but refusing to take it, would certainly deserve punishment.

In any case, there will be need for new rules and regulations, in social life and in the field of criminal law. As soon as it is scientifically established that some forms of criminal behaviour can be caused by certain genetic defects, is it no longer sufficient to punish a certain offence with a fixed penalty. One will have to consider for each individual case in how far the offence was genetically determined.

Criminal law considers that punishment ought to serve two purposes: retribution in the name of the citizenry and re-education of the offender. If it would turn out that a delinquent misbehaved purely as a result of bad genes, retribution would not really be an issue. Emphasis would shift to re-education, in the form of learning how to suppress one's bad urges.

Present practice is to keep dangerous delinquents who do not respond to therapeutic treatment forever locked away. Many of these individuals are probably precisely of the kind we discuss here: criminal due to faulty genes. When at some time it will become possible to make that diagnosis with absolute certainty, it will be an obvious choice to try gene therapy. So, verdicts may come to include compulsory gene therapy for mental disorders known to lead to criminal acts.

One always gets onto a slippery slope when one allows a certain kind of treatment in extreme cases. Why not also for cases that are slightly less severe? That would also be of benefit for society. Finally one would end up advocating that each young child should be screened for bad genes and in the case of defects should be forced to undergo gene therapy, if and when that would be available. It is impossible to predict how this kind of choices, ethically charged and very susceptible to ideological, religious and political framing, will turn out in various countries. It is to be feared that dictatorial regimes will tend to go far in enforcing treatment of imperfect subjects. They may see it as an even better way of showing the world how excellent their regime is than systematically doping their top athletes.

In order to outline dilemmas clearly, I oversimplified the relationship between a person's genetic make-up and his qualities. Only very few traits are determined one-on-one by a single gene. There is almost always a combination of several genes involved, each of which is at the same time playing a role in other gene combinations determining other qualities of the person. There further is the epigenetic environment, which determines in how far a gene is at all active. All this means that we are still far from being able to derive physical and mental qualities from a given DNA profile.

But once that time will come. One will first use that knowledge to correct genetic defects, but from there it will only be a small step towards experiments on improvement of the genetic constitution that we are endowed with as products of natural evolution. More about that in the last section of this chapter.

Cloning

A clone is a genetically identical copy of a biological organism. Identical twins are natural clones of each other. In the discussion about *artificial* cloning of humans we must differentiate between therapeutic and reproductive cloning. In the first case the aim is to grow material for medical use on the same person, for example for a skin transplant or for the replacement of a malfunctioning organ, without the risk of rejection by the immune system of the body. This is a very dynamic area of research, based on extensive experimenting with animals.

In the case of reproductive cloning the goal is to grow a complete individual from a single cell of the donor. One can think of essentially harmless reasons – maybe the person in question just wants to 'pass on' his or her genes but does not produce fertile sperm or egg cells – but reproductive cloning is also associated with the spectre of swarms of clones under command of an evil power. Maybe one day a mad dictator will make 100,000 copies of himself so that he can personally occupy every important position in his country. Or a leader of a religious sect could grow infinite numbers of devout followers, or an aggressive regime could install breeding factories of soldiers.

In short, we can imagine the most terrifying scenarios. The disturbing fact is that the technology required is in principle not very complicated. At present the rate of failure in the cloning of animals is still very high, but suppose that clever scientists at some major research institute find the recipe for successful cloning of any kind of mammal. It would then be quite a disaster if that recipe fell into the hands of a malignant power. Once they would know the procedures for successful cloning of humans, rogue states, mad sects or crime syndicates could find it much easier to go that road to world dominance than to start a programme for the production of nuclear weapons. The technology needed for clone breeding is far simpler than the vast machinery of nuclear high-tech. A few years ago a religious sect announced that it would soon be ready to present to the world the first human clone. That was bluff, or maybe something went wrong, but it may become true in the near future.

What is the procedure for cloning? The nucleus of a somatic cell of the donor – that means, any cell except egg or sperm cells, since cells involved in sexual reproduction only have half a set of DNA molecules – is inserted into an egg cell of the same biological species, from which the own nucleus has first been removed. In the case of reproductive cloning the egg cell will then be implanted in the womb of a surrogate mother. Cloning is still very inefficient. Most of the embryos already die in the womb and many of the ones that are born alive, show severe defects. (If the donor is a male, the egg cell used necessarily comes from a different person. In that case the clone will not be an exact genetic copy of the donor, because the body of any cell contains a few genes in its mitochondria, the energy producing units of a cell. Humans have 37 mitochondrial genes,[22] against more than 20,000 genes in the cell nucleus.)

If cloning is done for therapeutic reasons, the embryo stays in a test tube. When it has grown to about one hundred cells, stem cells are taken from it to start a cell culture. By manipulating the environment of the culture, one can stimulate the stem cells to form a specific type of body tissue. This tissue can be implanted in the donor without risk of rejection, because it is genetically identical to his own body cells.

Besides therapeutic and reproductive cloning, one can think of a third, at present still extremely unrealistic type of cloning: producing a substitute body when the original has been seriously damaged by an accident, an illness or old age. Of course, growing a clone body is just the first and simplest step if one aims at full replacement of a person. The difficulty of the next steps is what keeps this application of cloning in the very far future. The consciousness, the experiences of a lifetime, all acquired knowledge and skills, in short, the entire contents of the old brain would have to be merged into the new incarnation, either

22 A fault in mitochondrial DNA (mtDNA) may in very rare cases cause a grave disease. If it is known that a mother carries 'bad' mtDNA, that may be reason to implant the nucleus of her egg cell into the cell body of another woman. A child will then be born with DNA from three parents. This procedure was legalized in the United Kingdom early in 2015.

by physically transplanting the old brain into the new body or by somehow 'transferring' its contents, whatever that may mean.

The idea of the clone as a successor may anyway not be a good one, because the body cell that the clone is grown from, may somehow 'know' the age of its donor. According to the telomere theory of aging that is certainly the case. If the donor was already at a ripe age when the clone cell was taken from his body, the clone starts its life with short telomeres So, an old donor cannot expect his clone to outlive him for a long time. In principle, something could be done about that. One could take away some stem cells already in somebody's very early embryonic stage, keep them stored under conditions were hardly any cell division occurs, and much later use these cells to make a 'young' clone of the person in question. But these are things that will certainly not yet occur within the next hundred years.

In the United Nations there is as yet no consensus about a binding treaty on human cloning. Instead an non-binding declaration was accepted in 2005. It called for a ban on 'all forms of human cloning contrary to human dignity'. The Charter of Fundamental Rights of the European Union, proclaimed in the year 2000, explicitly bans reproductive cloning. In the USA, institutions subsidized by the federal government have provisionally been forbidden to perform any research on human cloning. Brazil, Canada and several other countries have followed the European example, but Asian countries show less scruples. There, cloned embryos are being used for all kinds of research.

Brain research

It is still quite a riddle how our brain works. Much progress has been made in the last decades, but the human brain is an extremely complicated object. Approximately one hundred billion nerve cells (neurons), almost without any visible relation between form and specific function, and many trillions of connections between these cells, running criss-cross in all directions: that is the bewildering physiological basis to be captured in a functional model if one aims to 'explain' or at least fully describe the workings of the brain.

Which big questions will be answered during the 21st century? It seems likely that it will become clear in great detail how our memory works. Already researchers agree more or less about what happens when a fact, a piece of knowledge, an experience, gets stored in the brain. Basically, when two neurons become activated simultaneously, the strength of signal transmission between these two can remain at a high level for a long time due to a change ('potentiation') in the synapse, the contact point where one neuron can transmit an electric pulse to the other by means of ion diffusion through a membrane. In short, recording of memories consists of durable strengthening of synaptic connections. The idea is that the recording of, for instance, the fact that Bamako is the capital of Mali, consists of the creation of a circuit of nerve cells that have been especially conditioned to signalling among each other.

Of course, that is far from explaining everything. How does the circuit for Bamako differ from that for Dakar, the capital of Senegal? Which procedure decides where the Bamako circuit will be created and how do supervising parts of my brain find that particular circuit back when I am asked for the capital of Mali? Which mechanism makes sure that the circuit where the fact 'capital' is stored, becomes connected to neural circuitry for the words 'Bamako' and 'Mali', or ' Dakar' and 'Senegal'? And how about the pictures of the Sahara and other associations that spontaneously arise when I think 'Bamako'? Have synaptic connections to circuits with Sahara pictures already been made when I first stored 'Bamako' as an item, or am I at the time of recall shopping around in my brain for associations? And so on, and so on. One easily comes up with tens of questions about storage and retrieval of simple memory items that the greatest experts in the field cannot yet answer.

But there is hope. Fast progress is to be expected now that researchers have the technical means to monitor the workings of the brain down to the level of individual neurons. The physiological structure of brain areas can now be mapped from cell to cell and the electric activity of groups of connected neurons can be followed. Moreover, much can be learned from experiments with neural networks in cultures of nerve cells. When one lets the network grow in such a way that it forms a flat structure, one can quite easily place an array of micro-electrodes against it in order to listen in to all activity or to selectively stimulate cells. A very important fact is that the computing power available for detailed simulation studies of brain functions is increasing fast.

When around the year 2050 the functioning of the memory is fully understood, a neurologist will be able to demonstrate to me how and where in my brain cells it is stored that Bamako is the capital of Mali. He will prove that my brain has stored this tiny bit of knowledge in a circuit of potentiated synapses at some spot in my brain. He will do that by erasing the potentiation with an electric pulse from some electrodes pointing to that particular spot in my brain, and by then asking me the name of the capital of Mali. I will have to admit that I don't know.

This little scene is of course a bit exaggerated. No overly enthusiastic neuroscientist will so easily be poking around with electrodes in my brain, and memorized facts are probably not stored in such a well-localized way as I pretended for the sake of the story, but still, the message is a serious one. Deep understanding of brain functions will lead to many beneficial applications, but in the wrong hands it can cause a lot of trouble.

The question of why we sleep and what happens in the sleeping brain will probably also be answered in the 21st century. Sleep research is bound to have points of contact with memory research, for it is already known that sleep plays an important role in the consolidation of newly acquired information. At the other side of the spectrum of brain puzzles, the side where breakthroughs are far less likely in the near future, the explanation of consciousness in terms of neural activity may be the most difficult nut to crack.

It is of little use to speculate which of the really difficult problems will in the coming hundred years be solved. Certain is in any case that progress will be large, for fundamental brain research is a hot topic, financially pampered in comparison with other branches of life science. In 2013 the European Union awarded funding of one billion euro to the Human Brain Project. This project aims to incorporate into a hierarchy of computer models everything known about the biochemistry of neurons, the exchange of signals between cells and the networks structures they form. This should within ten years result in a computer simulation of an entire brain. It is in two respects a very ambitious schedule. It is by no means certain whether all the necessary biochemical and physiological input data can be obtained in such a short time, and the kind of supercomputer needed to perform a realistic brain simulation does not yet exist on the market. Development of the computing machinery is in fact an integral part of the project.

The American government, not liking the idea of being overtaken by the Europeans, responded with the BRAIN project (Brain Research through Advancing Innovative Neurotechnologies). Its stated aims are about the same as those of the European project. The budget and the time scale are also quite similar. In reality the two sides will very likely agree on some form of work sharing. It is anyhow to be expected that their research programmes will become focused onto a number of core topics, rather than sticking to the grand idea of building a complete brain simulation, for which ten years and a few billion euros and dollars are certainly not enough. That would require at least ten times as much, in time and in funding. money.

Neurons can quite easily be connected with micro-electronic instruments. They form a happy marriage thanks to the fact that neural activity manifests itself as tiny electric currents and small voltage changes. The big difference is in speed. Electronic signals travel almost as fast as light (300,000 km per *second*), whereas neural signals from our brain propagate at the snail's pace of just 300 km per *hour*, so more than three million times slower. This is why an electronic calculator is so much faster than a person with pen and paper.

The method of signal detection by an array of micro-electrodes that pick up the electric activity of a single cell or a groups of cells, will become supplemented by other forms of neuro-electric contact. There are experiments with 'neurons on a chip', essentially a web of neurons immersed in an electrolytic solution, backed by a silicon chip with elements that sense the electric activity of each neuron in the web.

The coupling of signals of brain cells to electronic equipment clears the way to the control of instruments by thought power. Spectacular results have already been reached, primarily aimed at giving handicapped people control over the movements of artificial limbs, but the technology can just as well be used for the control of a robot arm, a vehicle or the cursor on a computer screen ('writing by thought').

Best known are the experiments that Kevin Warwick, an English scientist, carried out on himself. Already in 2012 he had a chip implanted in his arm, which was internally connected to his nerve system and externally connected wireless to a computer. With this he was able to manipulate an artificial hand and other objects.

In February 2013 an internet connection was realized between the brains of two rats that both were equipped with a neuro-electric interface. One animal was in the United States, the other in Brazil. They learned to stimulate each other to execute an action that would result in a reward. In other words, their brains had by trial and error learned to decode the electric signal carrying the thoughts of the internet partner. This implicates that humans should also be able to communicate from brain to brain with a kind of electronic Esperanto (and maybe also from man to rat and vice versa).

The modern hearing aid, the so-called cochlear implant, is the first commercially available piece of equipment that offers externally generated electric signals directly to the human brain. It essentially consists of a microphone plus a set of electrodes that transmit the electric pulses from the microphone to the hearing nerve in the cochlea. So, the aid takes over the functions of the outer, middle and inner ear and of the thousands of tiny hairs in the cochlea.

Visual prosthetics are still in an early experimental stage. They basically consist of a compact digital camera mounted on fake spectacles sending its signals to a neuro-electric interface. The interface may in principle be located anywhere along the path from eye to brain: at the visual neurons in the retina, at the optic nerve or even implanted in the visual cortex of the brain.

The technology of signal coupling between neurons and electronics can still be much improved. Most needed is further miniaturization, preferably down to a scale where it would become possible to introduce interfaces into the bloodstream and direct them through veins and capillary vessels to any desired location inside the brain. Surgical interventions would then no longer be necessary. Wireless communication with devices outside the skull will require a certain minimum size of the interface, which might make it necessary to launch it into the bloodstream as a building kit. Arrived at the place of destination, the parts would have to assemble themselves spontaneously, in the manner of self-assembling molecules, to form the interface complete with a transmitter and a receiver. It could run on energy beamed into the head electromagnetically, but it would be more elegant if the interface were a bio-electric hybrid feeding on nutrients in the bloodstream.

It seems likely that some kind of flexible technology for direct signal coupling to and from the brain will be developed already in the first half of the 21st century (but probably quite different from the fantasy I sketched). Of course, new and better techniques for scanning the brain from neuron to neuron in full activity will not only be important for applied research. They will be essential for a breakthrough in fundamental brain research. However, bio-electric implants sending their findings to the outside world will probably first be used in other, less delicate parts of the body and only later in the brain.

So far, I only discussed the use of neural interfaces in the context of prosthetics (sight, hearing, movement) for handicapped people. The aim in such cases is to let the user function as much as possible as a normal person. We did not yet touch upon the fact that, once an artificial eye has been developed for the visually handicapped, it will be quite easy to include extras like telescopic vision or sensitivity to infrared and ultraviolet light. For other kinds of prosthetics similar extensions beyond normal performance will be possible. Hearing aids could be made that detect ultrasound, possess adjustable frequency filters, provide better directional hearing, et cetera.

When one adds such features, one sets out on the path of improvement of man beyond his natural capacities. Technically that is not really different from building prosthetics for the handicapped, but there is a big difference in principle. The possibility that people may one day resort to self-improvement was in an earlier section already slightly touched upon in the context of genetic engineering. Now we have reached the point to confront the issue in its various aspects.

Better than nature

Natural evolution is blind. It results in life forms that are good enough to reproduce successfully, not more than that. Sometimes even a quite poorly designed species like the dodo passes the filter of natural selection thanks to a sheltered environment. On the other side of the spectrum, species that are much better equipped than is necessary for bare survival accidentally arise now and then, but generally life is a battle with imperfection.

That also applies to humans. How many unpleasant illnesses can we suffer, how easily do parts of our body get into disrepair, how terribly do we wear down when we get old, how we would love to be less slow, stupid, weak or ugly. And wouldn't it be nice if we could like bats navigate with echo location. And so on.

Soon man will have sufficient knowledge and know-how to have the option of freeing himself from a number of imperfections. He could do that in two ways: by replacing or supplementing his senses or other body parts with artificial components, or by genetic modification. We will discuss how likely it is that it would not just remain an option but become general practice, and which advantages and dangers that might bring.

Direct coupling of the brain to artificial devices was a topic in the previous section in the context of prosthetics for persons with a handicap. Much of the development work performed for that purpose will also be useful for connecting new kinds of sensory organs to the brains of healthy people. Echo location with ultra sound could for example be implemented as follows. In the sinus cavity one implants a small contraption that can emit a pulsed beam of ultrasound like a lighthouse and that measures how long it takes for echoes to return from different directions. Ultrasound emission could simply be switched on by the neuro-electric activity caused by the thought 'echo on'. The echo information could by software in the contraption be transposed into a visual simulation, that is, an information structure that could be fed directly into the visual cortex. One would actually 'see' the surroundings that one is scanning with one's ultrasound beam.

Sometime in the second half of the 21st century genetic technology and the functional understanding of the genome – which gene does what – will probably reach the point where genetic modification of humans is becoming a realistic option. A single intervention could then change more (and in a precisely defined way) than 100,000 years of natural evolution did (in random directions).

Natural evolution will then become to the human race just as passé as it has since centuries been for cattle and domestic animals. The enormous variety of canine races shows how much faster life forms can change once blind chance is replaced by purposeful intervention. Moreover, the diversity among dogs was achieved with the relatively primitive means of selective breeding. For humans, breeding programmes are ethically out of the question, but that is anyhow irrelevant, since genetic modification is a much more powerful technique.

In how far will people really want to improve themselves by cybernetic means (adding artificial body parts) or by genetic engineering? History shows that innovations which promise benefits in the way of improved well-being or competitive advantages stand a good chance of indeed coming into use, even if at the outset there are many ethical objections. I rather expect that the matter will until late in the 21st century remain in a kind of preliminary stage. When the general public becomes aware that technology has progressed to the point that much is possible, the first reaction in a democratic society will be to embark on a lengthy public debate about the question whether what is *possible* should also be *allowed*. This is natural and necessary, for a more important issue than changing ourselves is hard to find.

Public discussions seldom lead to a generally accepted conclusion, and on this ideologically and religiously charged subject that will certainly not be the case, but it does make people accustomed to the arguments pro and con. In a democracy, where one knows the value of compromise, there may then grow a general feeling that the issue at hand merits some degree of tolerance, similar to what happened in many countries on the subject of soft drugs and in some on euthanasia. Tolerating a practice or behaviour up to some limit without passing any formal legislation has the advantage that limits of tolerance can easily be adapted to shifts in public opinion. It may well be that man will indeed embark on the road of self-improvement in such a rather unspectacular way: starting with small cosmetic alterations, it will slowly become fashionable (and tolerated) to undergo more radical treatments. At some stages along the way, legislation will probably catch up, legalizing what has become practice.

The *reversible* attachment of accessories to sense organs, limbs or other body parts is already an old practice, although not to make man better but to correct deficiencies. Glasses and the wooden leg are examples. I only mention them here as a reminder that the question is not whether it is at all allowed to improve a person's performance by artificial means. It is a question of how and up to where. It is not a matter of black and white but of shades in gray.

The second kind of intervention, *irreversible* introduction of artificial devices, also occurs already in many forms. The artificial heart is an example, or the synthetic lens used in case of a cataract. Again, existing practice is limited to the correction of malfunction (except for some forms of cosmetic surgery). Discussion will start when healthy people wish to have devices implanted or hooked up to their bodies because they want to surpass their natural limits.

The possibility to directly control artificial body extensions with brain signals through implanted interfaces – which from the point of view of our brain would make them parts of our own body – is not of any real importance for that discussion. It makes little difference whether somebody moves a robot arm via a path of muscle power or directly by brain power. However, any direct connection between the brain and the outside world in the other direction – from the outside into the brain – deserves critical attention, because it could make the manipulation of people by other people possible. A brain chip working as a receiver of instructions from outside, can turn somebody into a slave. If the use of such dangerous hardware would ever be allowed, which might happen for medical reasons, the legislator should at least require that the device is designed in such a way that signals from outside can be received only if the wearer explicitly allows that by thought or action. Moreover, the explicit approval should be valid only for a short period. The receiving channel should close automatically before any malevolent sender could with a single message force the thoughts of the receiver to a permanent 'open'.

Interventions inside the brain need not be limited to the installation of interfaces that allow it to control external devices. One might choose to have certain brain functions executed by neuroprosthetics and not by the brain itself. This already happens sometimes for certain medical reasons, like loss of control over the bladder or to ease some cases of backache, but many more applications can be expected. Long before the brain is understood in its entirety, one will try to build devices that imitate parts of the brain that one thinks to know already. If such an artificial brain module performs successfully, one will want to let people with brain damage profit from it. If the product is developed further, up to a stage where it surpasses the biological original, people without brain damage will also want to have that module implanted, as an upgrade above their natural condition.

Product development, from first experiments with animals to availability for humans, will for many applications take many decades, but the beginnings are there already.

Researchers at Tel Aviv University have equipped rats with an artificial, simplified version of the cerebellum (the little brain) and connected it to the brain stem to control the movements of their legs. (Coordination of movements is the most important task of the cerebellum).

In principle, the visual cortex and other parts of the brain with specific functions and relatively transparent functional structures could also be candidates for replacement by an artificial counterpart, if the enormous problem of the innumerable connections with the rest of the brain could be solved. It seems unlikely that something of that magnitude will in the 21st century be tried on people, but experiments with animals will probably take place.

More drastic than adding or implanting artificial components, in reversible or irreversible ways, is the option of genetic modification of somatic cells (all body cells except egg and sperm cells). Until now, experiments have been limited to attempts to correct serious genetic defects, such as haemophilia and certain disorders of the immune system. These are ailments known to be caused by a single gene defect. In principle the method need not remain restricted to repairing faults in the genome. It also could be used to give healthy body cells extra capabilities.

Each type of intervention mentioned so far could give rise to protests, but they are actually rather harmless, in the sense that consequences are limited to the individual subjected to the intervention. A much more radical step would be too allow genetic modification of reproductive cells (sperm, unfertilized eggs or fertilized eggs), where the choice for fertilized egg cells is the most obvious. Many experiments have already been done with animals, including mammals, but as far as known not yet with humans.

The present procedure for genetic modification of an animal is as follows. First of all, the egg cell is fertilized in a test tube. Then the new gene that one wants to introduce into the genome of the fertilized egg is packed in such a way that it will have a reasonable chance to become correctly inserted in the DNA of the egg. This almost always means that the gene is first inserted into a harmless virus. Next, the treated virus is with a micro needle injected into the egg cell. A virus is not much more than a tiny package of DNA. It does not possess the biochemical machinery to reproduce itself. It must uses the services of a host. Therefore, viruses are specialists at invading the nucleus of a living cell and cunningly adding its own DNA to the genome of the host. In the case of the treated virus this means that it also adds the gene for which the whole exercise has been set up.

After the egg cell has divided into eight cells, one cell is taken off to test if the new gene has been properly inserted into the genome. If so, the modified pre-embryo is placed inside a womb. At the present stage of know-how it may take hundreds of trials for just one success, because the chance that the injected virus places the new gene correctly into the DNA of the egg cell is rather small. The small chance of success is already the first reason why the intervention is not yet being applied to humans. (Genetic *diagnostics* of a test tube embryo before implantation into a womb is available in a number of countries, although in most cases only to families known to run a high risk of an inherited disorder. The procedure then is that several test tube embryos are created. A test cell is taken from each of the embryos and if there is a healthy one among them, that one is implanted in the mother's womb).

Research into the genetic modification of humans is forbidden in many countries, or is very restrictively regulated, even if the intent would be to correct defective genes rather than to improve healthy ones. I do not expect that democratic countries will suddenly relax their rules sometime in the 21st century. It is more likely that rules will gradually be maintained less strictly, in the sense that exceptions will be granted or tolerated ever more often, in response to evolving public opinion.

It may start with allowing genetic correction of the fertilized egg if a couple can for some reason not produce a naturally healthy child. There will be demand for such a treatment as soon as technology has, by experimenting with animals, reached the point that the intervention has a high chance of success. Once this specific application of genetic modification has become generally accepted, people will ask for rules to be further relaxed. That will begin with requests to also allow genetic improvement for the purpose of cosmetic correction. The boundary between correction of a fault and pure embellishment is hazy whenever the word cosmetic is involved, so it could quite soon become a habit for parents to let features of their future child – hair colour, shape of the nose, body length – be adjusted to their taste.

Once a society has gotten this far, it is not a big step towards more radical interventions like the tuning of levels of hormone production in order to have a child with a pleasant character. Such a trend towards progressively more radical changes could lead to a shift in the genetic make-up of a population. At a certain moment the situation would be reached where parents would think to fail their child if they don't have the fertilized egg adjusted to be genetically healthier, more beautiful and more intelligent. That stage will probably not be reached in the time of one century, but the first step in that direction *will* likely be set in the 21*st* century.

It is not only because of ethical objections that genetic engineering of complex qualities like intelligence will not take place in the very near future. From data on identical twins it is known that intelligence is for eighty percent genetically determined, but it is far from clear which particular genes play a role, how these genes mutually interact, et cetera. Very few of our qualities and abilities are determined by just one single gene. In all other cases is will be very difficult to establish the recipe for genetic improvement.

And even if one finds the recipe for improving one quality, that may turn out to be harmful to a person's capacities in some other respects. For example, let us suppose that five genes play a role in dyslexia. It is very likely that each of these five would, in combination with still other genes, also be involved in other brain functions than word recognition. Optimization of a gene for word recognition will then have a high likelihood of harming the person's performance in those other areas. One will need a very detailed function map of gene combinations before genetic manipulation of complex human qualities can become a realistic option. In fact, it may quite often turn out that a certain correction is not feasible, because it would necessitate an almost endless chain of secondary corrections. The first, intended correction would have a negative effect on some other qualities of the person, so a second series of interventions would be needed, but these would harm still other aspects of his performance, et cetera.

Cybernetic modification – equipping somebody with artificial body parts – does not suffer from the disadvantage of everything being interwoven with everything. One does not become deaf from wearing visual prosthetics. Therefore, I expect that cybernetic means will generally be preferred if a certain upgrade of capacities can indeed be obtained that way. However, of many human qualities it is completely unclear how they could ever be improved by wearing some kind of device. Will there ever be a brain module for a sunny character?

Although it may be difficult to realize, there is nothing wrong with the idea of trying to improve imperfections of nature, genetically or with cybernetics, provided it really stays restricted to improvement. But how would a totalitarian regime exploit the possibility of genetic manipulation? By breeding obedient state servants, aggressive soldiers, tame citizens? An aggressive dictator might be even more tempted to offer his own people such 'advantages' of genetic engineering if at the same time democratic governments would stay very hesitant about allowing genetic improvement. He could breed what in his eyes would be a super race, capable of overrunning other countries. We can imagine all kinds of horror scenes from science fiction.

It is true in general that advances in science and technology make malicious regimes ever more dangerous, because they can misuse knowledge for ever larger evil. Genetic engineering is just one example. Nuclear physics, which led to the atomic bomb of North Korea, is an older case. What can one do against it? Banning scientific and technological knowledge is impossible. Once there, it cannot be made to disappear. A much firmer stance against malicious, unpredictable regimes than today also seems unrealistic. One would not gain much by, for instance, including in the charter of the United Nations a general obligation to intervene whenever a mad dictator grips power somewhere. It would just lead to a lot of bickering over the question which autocrat is dangerously mad and which one isn't. United action could only be expected after the United Nations were transformed into an organization similar to a world government, but that is not to be expected in the foreseeable future.

Does this mean there is no way out? No, it seems to me that there is a solution. It might precisely be genetic engineering that could prevent the nightmare scenario of mad leaders bringing death and ruin with state-of-the-art technology. We will return to this issue at the end of the book.

It spite of the dangers and problems of genetic modification, it seems likely that over five hundred or one thousand years people like us – susceptible to infectious illnesses and bad moods, suffering under frequent backaches et cetera– will no longer be the rule. Man as a purely natural creature will have been replaced by a genetically much improved version, optionally to be equipped further with cybernetic extras.

I do not believe in the Frankenstein scenario of super robots outflanking humans. The true believers in artificial intelligence underestimate man's advantage as a product of ruthless natural selection during a very long biological evolution. It is true that most people are not entirely satisfied with the constitution nature gave them, and would thus be interested in improvements, but even after a lot of genetic engineering they would still owe their capabilities for more than 99% to the genome evolution gave them. On the other hand, the hypothetic super machine has to be designed from scratch. That would not be so bad if it were approximately known how to do that, but so far nobody has the faintest idea of how to create artificial consciousness, just to mention one of the necessary ingredients.

A more distant future

Introduction

Up to this point I have limited myself to extrapolation of already existing lines of research in science and technology. In order to sketch innovations that could have significant impact on people's daily lives in the next hundred years, it was not necessary to speculate about discoveries or technologies outside the realm of present thinking. The year 2100, which served more or less as my time horizon, is as far from the present (2015) as the present is from the year 1930. In that year much of today's technological entourage already existed in some rudimentary material form or as ideas. That was my justification for assuming that I would, looking ahead over about the same time span, cover many of the most important developments if I could among the presently existing research lines identify the most promising ones from the point of view of practical applicability. But it is also true that information technology and molecular biology were nonexistent in 1930. Similarly, there will in 2100 probably be booming branches of technoscience that nobody can today have the slightest notion of.

The fundamental ignorance about radically new discoveries and inventions means that exploration of the future over a distance of significantly more than a century, can hardly pretend to be a serious undertaking. Nevertheless, I will make an attempt. I will assume that it may be possible to designate general areas of research where important breakthroughs in the course of the next centuries seem likely, if only because our present knowledge in those fields shows remarkable gaps. It is then still an open question *which* great novelties will in those general areas be discovered or invented, but I will venture to indicate possibilities, just to get away from too much vagueness. In order to have some focus, I choose a specific year. How might the world look like in 2500?

Before we speculate about futuristic technoscience, first the question must be answered whether mankind will by that time have entirely overcome the upheaval caused by the warming up of the biosphere. In the chapter on climate change I estimated that there would be no further warming from late in the 21st century onwards and I thought it unlikely that one would try to cool off the biosphere, even if the means to do it would be there. (Unless catastrophic melting of the Greenland ice cap would leave no choice, but I preferred to be an optimist on that account.) If this is how things will go, if the biosphere will indeed be left to cool down almost immeasurably slowly over many centuries, does that mean that the climate will have returned to a truly stable state by 2500? Maybe not entirely, because some of the measures taken in the course of the 21st century to slow down the temperature increase might have caused unforeseen side effects with a long aftermath, but essentially one should indeed have overcome the climate shock.

The efforts to keep the warming limited to just a few degrees Celsius will probably have resulted in knowledge that is generally useful for climate control. One may have found a way to slowly push the concentration of CO_2 in the atmosphere down by controlling the growth of algae in the oceans, or one may have learned to regulate the amount of sunlight absorbed in the biosphere by creating vast fields of clouds with variable reflective power. Even more powerful would be the capacity to push jet streams to higher or lower latitudes at will.[23] That would give control over the distribution of rainfall over the earth.

Before trying to look forward over five centuries, is it instructive to first look back over the same distance in time and to let it sink in that the next five centuries will bring even much more change than the past five did. Early in the 16*th* century, Luther and the Anabaptists were the initiators of the Reformation, Copernicus understood that the earth moves around the sun, Columbus had just discovered America without knowing it. The art of book printing was in Europe known since almost a century.[24] There were universities scattered over Europe, but the academic dish consisted of nothing more than theology, the study of law and erroneous medical theories.

Proper understanding of the laws of nature was lacking entirely. There was the know-how of craftsmen, farmers and seafarers, not much more. Cathedrals were built without any advance calculations. It

23 There are four jet streams, a polar and a subtropical one on each hemisphere. They are westerly winds at about 10 km altitude, always blowing at speeds of 100 km per hour or more. When one speaks of *the* jet stream, the polar stream of the northern hemisphere is meant. It blows over North America and the north of Europe.

24 That was not a first. Printing with Chinese characters had been invented in the 11*th* century already. Printing with single letters was practiced in Korea at the beginning of the 15*th* century.

was all done on the basis of experience and a keen eye. The study of science went no further than reading what the Ancients, the Greek and Roman philosophers, had proclaimed. Sometimes that had been right, more often very wrong, like Aristotle's statement that crawling vermin spontaneously arose from damp corners. There were some first attempts at experimental science, but within a context of alchemy or some other superstition.

Only in the 17*th* century did one begin to understand how the world really works. Newton formulated the laws of classical mechanics and optics. In the 18*th* century one came to a proper understanding of thermodynamics, which quickly found extremely useful application in the steam engine. This century saw the start of the continuous feedback cycle between science and technology, one pushing the other forward with ever new discoveries and inventions. The natural laws of electromagnetism were formulated in the 19*th* century. The 20*th* century became the era of man's exploration of nature at its smallest and largest scales: molecules, atoms and subatomic particles at one end, the structure of the universe at the other.

Anybody who has a feeling for what it means that we progressed in five hundred years from the know-how of carpenters and farmers to quantum physics and internet, from the abacus to differential equations, from the seven spheres of heaven to the Big Bang, must have an inkling of how much will change in the next five hundred years and how impossible it is to have any idea of the new directions of research that will be established. When I will try to point out some areas where important discoveries and inventions are to be expected, I am like the 16th-century craftsman who only knows muscle power and the power of wind and water and who is asked to make a forecast about power sources in the year 2000. The chance that he will mention anything resembling the combustion engine or the electric motor is nil.

Predictions in science and technology

Measured by the number of people involved in scientific and technological research and the technical and financial means they have at their disposal, progress in the next five hundred years should be at least a thousand times faster than in the last five hundred years. However, there is a saturation effect. The impressive scientific progress in the second half of the 19th century was the work of not more than a few dozen gentlemen-scientists, while much of today's research requires large research teams and huge budgets. The discovery of the Higgs particle in 2012-2013 was the result of a coordinated effort of thousands of scientists and engineers and an investment of billions of euros.

The width of the frontier of research, that is, the number of topics simultaneously being tackled anywhere in the world, may be a more reliable indicator of the speed at which technoscience will progress. Using that criterion, the estimate goes down a bit, but not by much. I would still expect science to advance several hundred times faster in the coming five hundred years than it did in the five hundred years behind us.

So, the world will be flooded with new concepts and innovations, some of which will change the life of the average citizen or broaden the intellectual scope of humanity. I will list developments and results that I think possible until the year 2500, on top of items that we already discussed when our time horizon still was around the year 2100. I start with the most probable and then descend to the less likely.

Very probable

1. The clean production of sufficient power will already around the year 2200 pose no problems anymore. Clean energy will be available in abundance from nuclear fusion reactors. They will supply at least 90% of the energy consumed worldwide. Only in very scarcely populated areas will this not be the best solution. Instead of building an extensive distribution network served by a single fusion plant, it will in such regions be better to stick to local energy production from some kind of 'free-range' source, maybe farms of sunlight converters backed by some type of energy storage for the nights. The last ruins of big wind parks will already have been demolished around the end of the 21st century. Fossil hydrocarbons will after the 21st century almost exclusively be used as raw materials for chemical industries, hardly as fuels for engines, power plants or heating.

2. Research in fundamental physics will in 2500 have led to a profound understanding of the microworld of elementary particles, the macroworld of cosmology

and the intimate connection between the two. It will be known since hundreds of years which phenomena are hiding behind what we, in the early 21st century, call 'black matter' and 'black energy'. The apparent incompatibility of quantum physics with the theory of general relativity will have been solved. In the process of solving old puzzles one will, however, have encountered new puzzles at a still deeper level, so there is no reason to assume that man will five centuries hence fully understand the universe.

3. Thanks to genetic engineering, people will in 2500 live much longer than today. Inheritable diseases will since centuries have disappeared. All kinds of improvements will have been made to the human body. That will not be restricted to purely physiological changes, such as better senses. Genetic modification may also have made mental disorders quite rare.

The extent to which manipulation of a person's mental state will be practiced, will depend less on what is technically possible than on rules and regulations. It is unpredictable how these will evolve with time. Norms and values can change enormously over a time span of five centuries. In an earlier chapter we discussed that one might in the history of mankind detect a trend towards increasingly peaceful, empathic behaviour. That gives hope that new technologies, such as the ability to modify the human genome, will generally be applied rather wisely, but neither must we forget that there have been numerous relapses into barbarism. It will be of utmost importance to prevent that tyrants of the type of Hitler or Stalin could abuse genetic technology for the creation of slavish subjects.

4. A similar danger is inherent to the cybernetic improvement of human capacities, which will probably become quite normal long before 2500. Whether people will be able to make their own choice from the available assortment of artificial body parts or will be forced to improve their natural capacities in a prescribed way, will entirely depend on the prevailing political system. In a democracy, everybody will be free within some reasonable range to choose for himself some useful or nice add-ons to the biological body, or may prefer to stay just natural, without any artificial parts. In a totalitarian state, subjects would likely be separated into distinct classes, with strict rules per class of how people should be fitted with auxiliary body parts.

Probable

1. The size of the world population will be regulated. Control of birth rates will have been introduced to offset the strongly increased life expectancy. It will be easy for authorities to enforce the rules, since fertilization in vitro will be general practice, given that the use of genetic diagnostics and modification will be the norm.

2. The weather can within certain limits be controlled. Maybe this will include nudging jet streams to particular positions, or inducing the appearance and disappearance of highs and lows, which would mean that one can control rainfall. It would also imply that tornados and hurricanes can be tamed at birth.

3. It will be fully understood how our brain works. There will be a complete bottom-up model of how interactions between neurons lead to consciousness and all other complex mental phenomena.

4. Once it is known how the human brain functions, one will be able to build copies, maybe in the form of electronic circuits, maybe in some other kind of hardware. That will be the moment when artificial intelligence is created. Since it will be possible to expand the amount of artificial brain capacity in the form of silicon chips of something similar almost without limit, the human original will from then on be surpassed, as long as it concerns activities that primarily are based on calculation power, logical reasoning and memory capacity.

The entity created in this way will be conscious, but will not behave like a human. A human being also has feelings and moods. These attributes depend on the interaction of brain activity with hormone production and other chemical processes in our body. I suspect that also qualities like initiative and creativity need a chemical impulse from the body. They are not phenomena that spontaneously well up in an isolated brain. An artificial being consisting of no more than a brain and sense organs will be like a tame sheep, without wishes or original ideas. That is very good, for it is just what we always wanted: intelligent servants who perfectly execute the tasks we give them, but who have no inclination at all to become masters themselves.

If one would want to create an artificially being that is really equivalent to humans, an entity with fantasy and surprising ideas, it would be necessary to copy the entire body with all its functions, not just neural circuitry of the brain. A chemical installation would have to be designed that does to the artificial brain what our body does to our biological brain. It is to be hoped that one will be wise enough not to embark on such a project. It would be extremely dangerous to have a superbly intelligent entity acting on its own initiative and according to its own tastes. Our understanding of good and evil, of what one should and shouldn't do, is the product of biological evolution. It is essentially the value system of a species living in extended family groups. An entity acting according to its own will and not sharing our heritage would almost certainly behave in a way that we find depraved and criminal.

Possible

1. The tame kind of intelligent machines discussed above could be built in the form of small spacecraft, suitable for reconnaissance missions to earthlike planets around the stars in our neighbourhood. Even if one only considers journeys to planets orbiting stars that are not more than ten or twenty light years away, the one-way trip still will take at least several centuries or maybe even some millennia, depending of the propulsion technique. Therefore, it is highly improbable that *results* of such an expedition would around the year 2500 already have been received. On the other hand, it is well possible that the technical means for missions outside our own solar system will become

available quite soon, maybe in one or two centuries from now, so we can imagine that in the year 2500 a fleet of spacecraft will already be on its way to planets circling other stars than our sun. In the next section, about the search for extraterrestrial life forms, we will examine this matter in some detail.

2. Man will in the year 2500 maybe have learned to keep somebody for almost unlimited duration in a state of metabolic standstill, similar to the natural phenomenon of hibernation, and to wake him up without lasting damage. All kinds of interesting applications would be possible. People suffering from an illness or injury for which there is as yet no treatment, could be kept in hibernation until a cure becomes available. Manned interstellar travel would in principle become possible. Journeys lasting centuries or millennia would not have to be left to machines for no other reason than the long duration.[25] (I think though that there are other arguments for not sending people on such expeditions, but that is for the next section.)

A much more interesting application of hibernation would be to offer every person the option to retreat into that state for any length of time. By doing so repeatedly, one could distribute one's life over several eras. It would lead to interesting situations. A grandmother waking up from a hibernation period of one century would be younger than a grandchild that had stayed active.

25 I prefer not to consider the idea of gigantic spaceships with crews that reproduce from generation to generation. It seems quite uncivilized to sentence many generations of as yet unborn children to a life of imprisonment on a spaceship.

One can of course argue about my classification of issues as (very) likely or just possible, and I probably left out subjects that other people would have put on top of their lists, but more important than the question which particular scientific and technological advances will stand out several centuries from now, is a different issue: how will forms of governments, politics and the administration of justice evolve over the centuries?

In an earlier chapter I have tried to address that question for the next hundred years, but it is impossible to say anything meaningful about the kind of public institutions that may over five hundred years exist in Europe or anywhere else. It is not difficult to imagine all kinds of possibilities, but how to separate the likely from the unlikely ones? One thing is clear though: the institutions of government and justice are in democratic societies by their nature slow, so that they will find it ever harder to update laws and regulations while society changes ever faster as a result of the increasing pace of progress in science and technology.

Given that democratic societies find themselves already now, early in the 21*st* century, severely handicapped by the inherent slowness of their system of checks and balances, it is doubtful whether there will in a few centuries still be a functioning form of government that by our present standards could still be called democratic. Parliamentary democracy might by then be a thing of the past, collapsed under the reality of ever faster technological innovations and replaced by the more streamlined procedures of (enlightened) dictatorship or by a government in the hands of a small circle of technocrats. I will return to this matter in the last chapter.

In search of extraterrestrial life

Space exploration was not on my list of core regions of technoscience in the 21st century.

It is an activity where many different disciplines of science and technology find application and which may be the basis for lots of interesting findings and innovations, but these belong to the various specific disciplines, not to a heading 'space research' or 'space exploration'. There also are fields of pure science, like astrophysics and cosmology, that profit very much from observational opportunities offered by space stations, but that certainly do not themselves belong under the heading of space research.

In my definition of space research, its one and only subject is the study of celestial bodies with spacecraft. I did not choose that as a main topic when we discussed the influence of scientific and technological results on society in the 21st century. Missions to study our sun, its planets, the moons of these planets and smaller constituents of our solar system have in the last decades of the 20th century produced a stream of interesting discoveries, and that is bound to be continued in the 21st century, but they will have no big impact on the spirit of the age.

Discovery of extraterrestrial life would be a different matter. That *would* make a lasting impression. However, the search for life in other places than the earth is a long-term project. It is not a topic that will have funding priority in the next hundred years. Authorities will in the coming decades discover that it is going to be very costly to keep global warming halfway under control. Projects without direct usefulness, such as the exploration of other celestial bodies, will suffer. It may well be that the budget that would in the USA still be left for space research, would for a large part be wasted on a return trip to Mars for two people.

Apart from satisfying patriotic sentiments it is a pointless undertaking to send people to Mars – robots can in space achieve much more than humans. The project 'Man on Mars' nevertheless appears to be developing into a pet project of successive American presidents. It could even be infectious. A manned race to Mars between China and the USA could become the next episode in a story of wasted effort and money that started with the race to the moon in the sixties of the 20*th* century between the USA and the Soviet Union.

Why would priority for Mars not benefit the search for extraterrestrial life? Hasn't Mars been considered the first place to look for it? That was indeed the state of affairs when conditions on Mars were still poorly known. Meanwhile, after a huge amount of data has been gathered by Mars orbiters and landers, it is evident that this planet is a very hostile environment for life.

What makes up a promising environment when one looks for life? To answer that question, we must first agree on a definition of life. It is often said that it is very difficult to define life in general terms, but it is not really hard if one sticks to measurable characteristics and stays away from philosophical contemplations. Almost everybody should more or less agree with the following: a living creature is self-regulating in all its functions, extracts energy and materials for growth and maintenance of its body from its environment, reacts to stimuli, can adapt to changes in its environment and reproduces itself.

This list shows that any life form anywhere in the universe must in its molecular structure be quite similar to creatures on earth. Self-organizing macromolecules, various sorts of proteins, that kind of biochemical kitchen based on carbon, hydrogen, nitrogen and oxygen must be the starting point, because they are the only ingredients with which the characteristic qualities of life can be realized.

Water is best suited for such complex molecules to float around in, for very long times, without damage to their structure until some of them accidentally fuse to form an assembly of even higher complexity, et cetera, until after ages primitive life comes into existence. There are other liquids that could replace water, so there might be life at places that are too cold or too hot for liquid water, but it shouldn't be very cold, else chemical reactions would proceed very slowly. Too hot would be even worse, for macromolecules are not stable at elevated temperatures.

So how about the Martian environment? The surface of Mars is bone-dry and frozen. Because there is almost no atmosphere, it is exposed to a high dose of ultraviolet radiation from the sun. Proteins or other macromolecules would quickly be destroyed under this bombardment. Laboratory tests with microbes have indeed shown that the UV dose on Mars is devastating. Even worse is that Mars, in contrast to earth, has no magnetic field, which means that there is no umbrella against cosmic rays. It has been calculated that life on Mars is for that reason impossible down to a depth of ten meters under the surface. The question then is whether there is liquid water somewhere at larger depths, and if so, whether a living organism could find there some source of energy. The chances seem small.

Bodies in our solar system where conditions appear to be more favourable are the moons Europa, Ganymede and Callisto of the biggest planet Jupiter and the moons Titan and Enceladus of the ring planet Saturn. Europa seems the most attractive of the three Jupiter moons. It is covered with an ice crust, several kilometres thick, on top of a water ocean. Titan, the biggest moon of Saturn, has a thick atmosphere above a landscape that looks somewhat like earth, with dunes, rivers, lakes and oceans. However, what flows there is not water but liquid methane and ethane. Enceladus is similar to Europa: entirely covered with ice and below that probably an ocean of liquid water.

Europa

The tides of Jupiter act like a kneading machine on the solid core of Europa. The heat this generates keeps the water under the ice crust liquid. The ocean is probably about a hundred kilometres deep. The bottom would be the place to hunt for life. The gravitational force exerted on the core of Europa by Jupiter must lead to processes similar to the shifting of tectonic plates on earth, resulting in volcanic eruptions along rifts on the ocean floor. On earth, such 'hot spots' are islands of life in the depths of the ocean. Life on earth may well have started there.

Most data on Europa were gathered by the Galileo probe, launched way back in 1989. The next close look at Europa is on the programme of the Jupiter Icy Moon Explorer of, very appropriately, the European Space Agency. The launch is planned for 2022. An actual landing on Europa is not foreseen.

Titan

Titan has a thick atmosphere of mostly nitrogen with a dense smog of hydrocarbons. We know how the surface looks thanks to the European-American Cassini-Huygens probe. This space craft reached Titan in July 2004. Half a year later the lander Huygens descended through the atmosphere and landed on the surface of Titan. Pictures taken with radar and optical cameras show lakes of liquid methane or ethane in the polar regions of Titan.

Clouds of methane drift in the air, there is rain and wind. In short, the conditions are almost earthlike, except that the role of water is on Titan taken over by methane. The high concentration of that gas in the atmosphere causes a strong greenhouse effect. It is opposed by the fact that the thick smog scatters much of the incident sunlight back into space, but the net effect is that Titan is warmer than would have been the case without an atmosphere. Still, it is not hot on the surface of Titan. It is only about -180 °C.

Chemistry is slow at such temperatures, but NASA nevertheless reported in 2013 on the basis of simulations of Titans atmosphere that complex organic molecules could form there. Life is maybe possible in the methane lakes. Life forms might inhale hydrogen instead of oxygen and exhale methane instead of carbon dioxide.

No space organization has a further exploration of Titan in its present plans. The chance for any really close look at the moons of Jupiter and Saturn seems in the 21st century small for the two reasons I already mentioned: a good one (the costs of the climate crisis) and a poor one (priority for an extremely expensive manned mission to Mars). But anyhow, although the exploration of our solar system may proceed rather slowly and hesitantly, it will certainly have been investigated into its last corners *before* a first journey to another star is undertaken, because that will have to wait for at least two or three centuries. Present technical know-how falls miserably short of what is needed for space flight outside our own solar system.

The motivation to go out and look for life elsewhere is not lacking. Since a couple of years it is no longer merely expected but proven that many earthlike planets exist in our galaxy. Our instinctive curiosity about life forms, intelligent or not, on such planets is large. The discovery of primitive extraterrestrial life within our own solar system would be an extra stimulant, because it would be definite proof that life outside earth is possible, but also without such an additional push there will be general support for launching space probes to planets around other stars in search of life forms.

The first observation of planets around another star were based on the fact that a star and its planets must move around a common centre of gravity that does not coincide with the centre of the star. The result is that the star seems to wobble slowly to and fro. It is a tiny effect, only detectable if the star has very heavy planets. Another drawback of the method is that the observed wobble is the combined effect of all planets of that star. It is practically impossible to decompose it into contributions of individual planets.

Higher sensitivity can be reached by searching with extreme precision for variations in the light intensity of the star. If one periodically observes a little dip, one may conclude that a planet of that star is crossing the line of sight between the star and the earth every time it completes a revolution. (Provided we know for sure that it is not a double star where the partners cover each other periodically). Of course, the chance is small that the orbit of a planet has precisely the correct orientation to betray itself in this way. It amounts to about 1 in 200. The big advantage of this method is that it is sufficiently sensitive to detect planets of the size of the earth. In a few very favourable cases it will even be possible to measure the composition of the atmosphere of the planet passing in front of the star by spectroscopic analysis of light that, on its way from the star to the earth, passed through the atmosphere of the planet.

These are extremely delicate measurements that can only be done outside the earth atmosphere. Late in 2013 it was estimated on the basis of observations with the Kepler space observatory of NASA that our galaxy (the Milky Way) contains roughly 11 billion planets of a size similar to the earth *and* circling a star similar to our sun at a 'distance of habitability'. In the list of 2013, the nearest planet in this class orbits a star at 12.5 light years distance from us. In this context, 'habitable' means that the planet receives the right amount of light from its star to have a surface temperature where water can exist as a liquid, provided there is a sufficiently dense atmosphere.

Let us, as a thought experiment, assume that further scrutiny will show that his planet at 12.5 light years indeed is the nearest of all possibly life carrying planets. For an astronomer 12.5 light years is close by. It is in our nearest neighbourhood, only three times as far as Proxima Centauri, the nearest neighbour of our sun. However, for present space technology it is an almost insurmountable distance. A much more powerful method of propulsion has to be developed before it makes any sense to send spacecraft on interstellar journeys.

Just to give an idea of the enormity of the problem, the space probe Voyager 1 is now leaving our solar system at a speed of about 60,000 km per hour. By earthly standards that is fast, but if Voyager were on course to Proxima Centauri – which it isn't – it would only arrive after 75,000 years. So, at arrival its launch would be just as far back in time as Neanderthal people are for us.

If one does not want the journey to the planet at 12.5 light years to last longer than say a century (eight times as long as the 12.5 years a light ray needs), the space probe must travel at one eighth of the speed of light. That amounts to 37,500 km per *second*. At that speed it takes only ten seconds to get from earth to the moon. If one would want to send people along, then a space ship of tens of thousands of kilograms would be needed at the present stage of technology. A few centuries of progress in propulsion technique, power generation, construction materials and food cultures still wouldn't reduce that to less than a thousand kilograms. The energy needed to accelerate a mass of 1000 kg to one eighth of the light speed is 200 billion kilowatt-hour, which corresponds to the energy production of a medium-sized power plant during fifty years of continuous operation. The conclusion is clear: there is no place for humans in interstellar travel.

The energy cost of accelerating a space craft to cruising speed is proportional to its mass, so a robotic microprobe of a mass of 100 grams would 'only' need 20 million kilowatt-hour – attention: millions instead of billions, just two days worth of running a typical power plant – to be accelerated to the speed needed for reaching our star at 12.5 light years in one century. Energy savings are even easier to realize if one is willing to be more patient, but that has its limits. Energy of movement (kinetic energy) is proportional to the square of velocity, so the energy requirement drops by a factor of four if the journey is allowed to last two centuries instead of one.

There are three ways for a spacecraft to obtain its kinetic energy: from its own rocket propulsion, by sailing on a favourable 'wind' in space or by being launched like a projectile. Let us investigate their potential for travel to other stars.

The rocket is a reaction engine: backward ejection of mass at high speed gives the spacecraft a push forward. Ejection by the explosive force of combustion or some other chemical reaction suffers from a rather low ejection velocity. That must be compensated by pushing out a lot of material, which means that the spacecraft has to start its journey with an enormous weight if it wants to reach a high cruising speed. Therefore, the classic rocket engine is not an option for interstellar travel. Much higher ejection speed is reached with an ion engine, in which ionized gas atoms are accelerated electrically. Such engines are often used as steering rockets to manoeuvre earth satellites precisely into position, but they are useless if one needs high propulsive power, because the engine can expel only a small amount of mass per second..

Nuclear energy in principle offers the best chance for interstellar space travel by rocket propulsion. It can heat ejection gas to extremely high temperature, which means that the ejection velocity will be high. The first research in this direction was started already half a century ago. NASA currently has a study project on a rocket drive based on pulsed nuclear fusion. The nice point about using nuclear fusion is that the fuel, hydrogen gas, is floating around everywhere in space, so one can refuel anytime. But because the advent of controlled nuclear fusion – in contrast to using it as a bomb – is not to be expected before the middle of the 21st century, a fusion rocket could certainly not be built before the 22nd century, if at all.

How about sailing on a space wind of radiation or particles? It is an elegant idea, but the propulsion power is very small, and that only in the neighbourhood of the sun. Further away, the wind is far too weak. So, totally useless for interstellar travel, but possibly interesting for missions inside our solar system when there is no hurry. The first and up to now only example is the Japanese probe Ikaros, launched in 2010. It carries a big sail. The pressure of sunlight on the sail produces a small but continuous acceleration.

The sun not only emits visible light and other electromagnetic radiation, but it also ejects winds of electrically charged particles. They cause the polar lights. This particle wind could be used by enveloping a spacecraft with an electric or magnetic field that bends the trajectories of approaching particles. The field, and with it the spacecraft, then experiences a reaction force that accelerates the craft. The European Union has awarded a small subsidy for a feasibility study of this idea of 'electric sailing' on the solar wind.

The third principle of propulsion was already described by Jules Verne in 1865 in his science fiction novel *From the earth to the moon*. He had a spacecraft shot to the moon by a canon. The canon could be replaced by an electromagnetic catapult. However, such an installation for bringing a space probe with a single blow to cruising speed could in reality never be used on earth, because the probe would evaporate in the atmosphere like a meteor. The moon would be a better place. It has no atmosphere.

Acceleration of atomic and subatomic particles to speeds close to the velocity of light with electromagnetic forces is a well-known technique in research laboratories. It is still utterly impossible to do something similar with macroscopic objects, but let us assume, just as a thought experiment, that at some time the technology will exist to accelerate a small space probe to 30,000 km per second (one tenth of the velocity of light c). If one would build the accelerating track as a ring around the moon, the probe could be accelerated at a leisurely pace, just a small push at each revolution. Would that work? The answer is no, because in the end it would still be an extremely violent process, because of the high final speed required. In order to keep the probe on its circular track at a velocity of almost 0.1 *c*, electromagnets along the track would have to exert an enormous sideways force on it. I calculate that the sideways acceleration felt by the probe would be about fifty times larger than the most extreme acceleration to which we submit any material nowadays. (That happens in centrifuges for the enrichment of uranium.) It is extremely unlikely that the instruments in a space probe could survive such a mistreatment.

It would not help to build a straight acceleration track instead of a ring. The longest straight track on the moon (3480 km) would be a tunnel right through it. Bringing an object to a velocity of 0.1 *c* by constant acceleration over that track length requires a forward acceleration that is indeed smaller than the worst sideways acceleration experienced in the circular setup, but the gain is not enough, only a factor of four.

The Verne method is just not suitable for the high cruising speed required for interstellar travel. On the other hand, a catapult facility on the moon could be a realistic possibility for journeys within our solar system, assuming that the necessary acceleration technology could be developed and built at acceptable cost. At a given length of the catapult track, the required acceleration is proportional to the square of the speed that one wants the space probe to have. For travel in our solar system a speed of 30 km/sec is certainly sufficient. That is a factor thousand less than the interstellar speed I used above. In the required acceleration that counts squared, so there we gain a factor of one million. Launching a space probe with a speed of 30 km/sec (more than 100,000 km per hour) from a catapult track on the moon of say 1000 km length would involve an acceleration that is 45 times the gravity acceleration we experience on the surface of the earth. It would still be killing for people, but a well-constructed unmanned craft could endure it.

We have seen that there is as yet no propulsion technology that would allow a trip to a star nearby in a travel time of one or two centuries. Something entirely new is needed to make that possible. It is tempting to speculate about a route 'outside space', in the way that entangled quantum particles don't care about spatial separation, but that is pure science fiction with the emphasis on fiction.

The idea of exploiting phenomena in space in the manner of sun sailor Ikaros is in principle more realistic, but the question is what and how. Wait for a space storm of particles after a supernova explosion? Find a way to exploit the energy of the vacuum? Field theories of fundamental interactions describe everything happening in the universe in terms of fields that permeate space. From this perspective, the vacuum is more than just empty space. Instead, it is the state of lowest energy of the electromagnetic field, the Higgs field and any other field. According to quantum physics that lowest energy is *not* equal to zero. It is, however, unclear if one could profit from that in any practical way, and even if it were possible in principle, mankind might still need thousands of years to reach the stage of sophistication needed for that trick.

For the less distant future a marriage of rocket technology with nuclear fusion may be the best bet. That is not for tomorrow either, but it seems nevertheless likely that in the year 2500 space probes will be on their way to other stars, either with such a tuned-up rocket engine, or with a hitherto unknown method of propulsion.

Grand perspectives open up when it becomes possible to travel at speeds close to the speed of light. That will certainly not happen within five hundred or even thousand years, but let us for once think ahead much further. People travelling that fast can during their life time repeatedly hop from star to star. Essential is the phrasing '*their* life time'. For a person moving through the galaxy at almost light speed, time runs slower than for the people who stay at home. His spacecraft, his body and everything on board experiences the relativistic phenomenon of the stretching of time.

According to the people at home, a vessel travelling at 99.5% of light speed to a star one hundred light years away will reach that star after a little bit more than one hundred years, but for the traveller the journey only takes about ten years. Everything in the superfast spacecraft, from the clock speed of the board computers to the aging of the body cells of the traveller, happens ten times slower than on earth, just like a slow-motion movie. At least, that is how the people on earth see it. The space traveller and his instruments do not experience any slow-down. They see time passing by just normally. But then how do they explain that, according to their board clocks, they only needed ten years to reach a star that was hundred light years away? Wouldn't that mean that they had travelling ten times faster than light, which is known to be impossible? No, the answer is that, from the point of view of the fast traveller, the distance from earth to that star is only ten light years instead of hundred.

In summary, the way in which the relativistic effect of very fast travel manifests itself depends on the state of movement of the observer. According to people at home, things on the fast ship happen in slow-motion. According to the traveller, that is not so. For him the distance to be travelled has shrunk. This state of affairs shows that space and time are not a universal background, a single fixed theatre in which all action takes place. In reality, each state of movement has its own version of space and time.

In an earlier chapter I mentioned that instruments for GPS-navigation must take into account the relativistic difference between the 'rhythms of time' in a satellite broadcasting the signals and in a receiver on the surface of the earth. In that case it is a tiny difference, which only becomes important because mismatch accumulates as time goes by. Really big effects only occur at velocities close to the speed of light (and in extreme regions of space like black holes). In a space ship travelling at half the speed of light, time on board ticks by only 13% slower than on earth. Only in the last percentages of closing in on the light speed does the relativistic effect of time stretching – or distance shrinking, depending on one's state of movement – take on dramatic size.

What would be the benefit of travelling almost as fast as light, and what would it cost? At the present state of science, the balance is quite negative. To start with, as long as no humans come along, there is no real gain at all, because time stretch in an unmanned probe is unimportant, assuming that the instruments will have been constructed to last long or to be self-repairing. The only gain is that a star at *x* light years distance will, according to the time reckoning of people on earth, be reached in only slightly more than *x* years. That is indeed better than a probe moving with a tenth of the light speed, which arrives only after *10 x* years, but it is a gain without a relativistic bonus. On the side of costs there *is* a heavy relativistic effect: it takes enormously much more energy to accelerate a massive object to near the speed of light. This phenomenon and the effect of time stretch become noticeable at the same speed and increase equally fast upon further acceleration. In other words, space travellers will in a faraway future only be able to enjoy the advantage of time stretch if their ship in some way has almost unlimited propulsion power.

Matters of concern

Many of the issues we discussed are surrounded by a haze of uncertainty. That can't be helped. However, little doubt exists about some general trends: the growth of the world population will soon stop, globalization will continue in many fields of human activity, the fight against climate change will to some extent be lost, science and technology will bring new products and ideas at ever increasing rate, man will choose to improve himself by genetic and cybernetic means.

What is absolutely certain, is that the speed of technological innovation will in some instances lead to situations of crisis. I have pointed out several cases. Three of them seem so important to me that I want to consider them again in this final chapter. It concerns the need of an overhaul of the system of parliamentary democracy, the fact that a world government will soon be desirable but in practice impossible, and finally the most important issue: the risk of catastrophic events of human origin. May humanity kill itself with its own inventions?

Renovation of the democratic system

Improvement of the standard of living of the average citizen has always come from science and technology. In the 21st century that will be especially true, because the world population will in the first half of the century still be growing, while at the same time costly measures have to be taken to keep the warming of the earth limited to a few degrees. Under such difficult circumstances one will have to rely strongly on technological innovation if one still wants to realize some progress in the conditions of life, particularly in the poorest regions of the earth. It will, therefore, be important that governments and politicians develop more affinity with technoscience than presently is the case, both in developed and underdeveloped countries. A good grasp of its potential benefits and dangers counts more than occasional willingness to support some kind of R&D for political reasons.

For a proper understanding among political decision makers of how technology is changing society, more positions in government would have to be occupied by persons educated in science or technology. Of course, such people should also possess wisdom and managerial aptitude – the qualities the present mainstream administrator adorns himself with – but that should come on top of solid professional qualifications. The habitual failure at a grand scale of government projects in infrastructure and informatics is just one (but very visible) example of the fact that one should not take decisions on matters that one doesn't oneself understand. People with power to decide often live under the delusion that it is not necessary to be themselves knowledgeable about the matters they decide. They think to possess sufficient wisdom, intuitive insight and experience to be able to rate the views of experts at their true value on any subject. In truth that is impossible for a layman.

It would be a valuable step forward if in future top positions in government and public administration would be reserved for people possessing wisdom *and* knowledge, but that alone is not enough to use the fruits of technoscience as well as possible. It is not only a matter of having the right people at the right places. Equally important is the framework of decision making and public control within which they will operate.

Democracy in its purest form – every vote has the same weight and the majority wins – unfortunately seems to be rather poorly suited for dealing with science and technology. Research areas like nanotechnology, artificial intelligence and genetic technology will bring discoveries that can radically change the life of the average citizen. The future of mankind will to a large extent depend on decisions about allowing or forbidding certain techniques, but only few voters will understand such matters sufficiently well to be able to make up their own mind. Without changes to the way our parliamentary democracy works, the result would be that a large part of the population becomes a voting mob for demagogues.

It may in principle be sensible to leave decision making about crucial but complex questions to a body of wise scientists and scholars, but it sounds quite unattractive to let a 'council of the wise' decide about matters of great importance without any democratic control. Or one might add an extra layer to our parliamentary system, only for important affairs that are beyond the grasp of the average citizen. In this layer there would be no universal right to vote or to be a candidate for election. Only citizens who would be found, by certain criteria, to be sufficiently judicious, could vote on candidates that would be selected by a similar procedure. The members of the extra layer of parliament would be entrusted with decisions on 'difficult affairs'.

Such a scheme appears contrived and unsatisfactory, but I just wanted to show that it will be hard to find a decent solution for the fundamental dilemma of a democratic system where everybody's vote counts, even if the voter doesn't have the faintest idea what he is voting for or against. It is in principle the most civilized form of government, but also a dangerous system when many voters do not understand the issues at hand. Unsavoury types with populist slogans may then get themselves voted into power.

A less fundamental problem, but with considerable practical consequences, is the slow pace of legislation in a democracy. A political system founded on the separation of powers, public debate and rivalry between political parties will never be able to compete with autocratic systems on the issue of speed and shouldn't even want to, but it is a fact that it will become ever more important to act as quickly as possible within the limitations of the system. Already, law makers are lagging ever further behind actual developments, especially in the field of internet use and digital data storage. If procedures of law making are not streamlined, the increasing pace of technical innovation will within a few decades create a state of technological anarchy.

Waiting for a world government

Why is so little accomplished at the annual conferences about climate change? That is because national leaders primarily defend their narrow national interests. So far, only some European countries have shown themselves able to think further. In world politics the nation state still takes precedence. In the past centuries the concept of the sovereign state has been useful for the maintenance of international order, but meanwhile globalization has progressed so far that almost every big problem requires a joint approach. On the subject of climate change that is acknowledged in principle by every government, but once a plan of action collides with real or imagined national interests, the narrow style of thinking immediately takes over.

The reasons for that behaviour are obvious. The leaders of democratic countries have been chosen on the promise of optimally serving the interests of the voters. Dictatorial rulers defend the interests of themselves and their cliques. The latter will come to a half-hearted turn if they estimate that joining a global effort at problem solving is necessary for the survival of their regime. Things are more subtle in democracies. There, many politicians are well aware that the unassailable nation state is an outdated concept, but only few of them have the courage to tell it to their voters. They fear that such a message will cost them votes, and they may unfortunately be right. So, a member of parliament may for instance choose *not* to support a common approach against global warming that he knows to be right, but that would in the short term have a negative effect on the economy of the district he represents.

Most citizens have no idea how strongly the world is interwoven already. The communality of problems has outpaced the communality of minds. Cultural differences are still too big to expect widespread support for the idea of a supranational world government within one or two centuries. If the trend towards globalization keeps doing its work, one day the world will be ripe for some kind of common government system, but until that day one will have to tackle common problems of the magnitude of the climate crisis by collaborating in the framework of multilateral institutions, of the type of daughter organizations of the United Nations. Those are far from perfect, but institutions like the World Health Organization, the International Labour Organization and the International Atomic Energy Agency have proven that they can do important work, as long as one does not politicize every issue.

To finish this section on an optimistic note, let us remember that in 2005 an important step was taken, at least on paper, to break the doctrine of unconditional national sovereignty.

In that year the General Assembly of the United Nations declared that the international community has the responsibility to protect citizens who are suppressed by their own government.

Man-made disasters

People are by nature interested in the survival of their offspring. That is part of our genetic heritage. It is the reason why we care about the effects of our actions far beyond our own time of life. Therefore, the most important question an exploration of the future should try to answer, is how mankind should behave in order to maximize its chance of survival into the distant future. I will not try to come up with rules of behaviour for more or less normal times. I only want to investigate if there is a strategy to prevent grave disasters inflicted by man upon himself.

As technology progresses, the consequences of acts of stupidity and malevolence can become ever more grave. In the near future a series of political miscalculations could for example lead to a situation where the Taliban get their hands on Pakistan's nuclear arsenal. Or clumsy moves in a territorial or economic conflict could escalate into a world war with nuclear weapons, without either side having intended that outcome. In the more distant future, when it will be possible to genetically tailor micro-organisms to one's desire, biological terrorism may become the worst threat. If a new kind of pathogenic agent is distributed to which our immune system is defenceless, that will be the end of mankind.

Catastrophes of human origin are nearly always the result of miscalculation. That does not only hold for blunders caused by incompetence or inattentiveness, but also for the mischief perpetrated by malicious types, for such villains take advantage of the miscalculations of their opponents. There is no doubt that Hitler started World War II deliberately, not by mistake, but he would not have come to power if the alliance of nations that had besieged Germany in World War I hadn't made the error of imposing extremely heavy burdens on the country in the Treaty of Versailles of 1919.

There is no standard recipe for preventing fatal errors. The advice to rethink at least twice before taking any drastic action is quite unrealistic as long as our society is not organized like a colony of ants. If everybody would act as if he were part of a single organism, one could indeed institute the rule that nothing important would be done without elaborate pre-evaluation of the possible consequences. Reality is different. It is in our nature as a clan animal to split into competing groups. Nations, companies, sports clubs, in every area of human activity organizations are in competition with each other. That fact has been a driving force behind progress in knowledge and know-how, but it also means that decisions are often taken under some form of time pressure. There is always the fear that others may overtake you.

Would it be advantageous for mankind – as a species, ignoring individual well-being – if people learned to behave like ants? Suppose that at some future time people would, by genetic engineering or brainwashing, be conditioned to march in step. Would that improve the chance of survival of the species? I don't think so. The risk of disastrous decisions being taken would indeed be reduced if one could afford to deliberate at length on everything, but such a habit could become a serious handicap in a real emergency. A more serious problem of the anthill model is that the entire colony may after long deliberation decide to march into the wrong direction. In the competition model, mistakes by one party can usually be corrected quickly enough by another to prevent catastrophe, but if everybody marches to the same tune, all will tumble down the same cliff. The greatest danger of marching in step is that it would make it much easier for a malicious clique to take over the power over the world, which is the ultimate nightmare. (This does not mean that the idea of a world government is wrong. The danger is in the synchronization of minds, not in uniting the world under one government. A world government is no more threatening than a government at the level of a nation state, provided there is a system of controls and balances like in present-day western democracies.)

I am optimistic in the sense that I do not consider the anthill scenario as a likely future of mankind, but that forces me to assume that it will remain true that important decisions are often taken in a hurry. Hurry will still not be a good adviser, for it is unlikely that man can soon blindly trust the advice of supercomputers or counsellors with artificially improved brain power. How then can we avoid that human stupidity or maliciousness will at some time lead to a enormous catastrophe? As long as people are as they are by nature, there remains in principle one way to exclude man-made disaster, but it is very unlikely that it would be successful. One would have to give to an institution that itself is fully trustworthy the mandate and the means to operate some kind of worldwide warning system against unwanted activities and the power to stop them. In other words, an incorruptible world police that would be above the law in every country. It is hard to believe that an institution with so much power would be able to maintain its integrity. It is even more improbable that nations unwilling to surrender any degree of sovereignty to a world *gov-*

ernment would be willing to do so to a kind of world *police.*

And so it seems impossible to prevent that at some time a mad dictator will blow up the world, a fanatic leader of a sect will exterminate humanity or a stupid autocrat will in his delusions take a disastrous decision. One should comprehend that this is a matter of all or nothing. If the chance that a dangerous madman may strike is not reduce to nil, it will happen one day. For instance, assume that human nature and the structure of our society are such that in any given year we run a chance of 1 in 10,000 that a madman will indeed perform an act of total destruction. Mankind then runs a risk of ten percent per millennium of becoming extinct. That sort of risk is maybe acceptable for a giant tsunami swallowing a million people, but not for the extermination of humanity.

Having established that an apocalyptic event can never be completely excluded as long as people are as they are by nature, only one solution remains. Mankind should decide to change itself in such a way that dangerous madmen don't exist anymore. This seems to be the inescapable conclusion. Man's technology will become so powerful that he becomes a danger to himself if he stays as he is today. As an act of self-protection, traits like hunger for power and inclination to violence and deceit will have to be eliminated. That will in a not very distant future be possible with genetic engineering supplemented by hormone treatment.

It is not a pleasant prospect. Still, a global campaign of genetic modification could be possible in the manner of the worldwide smallpox vaccination in the 20*th* century *if* the necessity were generally understood, by authorities and by the people. However, such understanding is not likely. There probably would be much aversion against enforced genetic modification and against character tests that could label somebody as a dangerous individual in need of supplemental hormone treatment. Presumably, some terrible man-made catastrophes would have to happen before the programme would receive general support, and even then there would remain strong religious resistance in several corners of the earth.

As long as the world population will not accept in large majority that man has to change himself in order to prevent his own annihilation, one might try to approximate the idea of an incorruptible world police, but that will never result in truly effective control. As long as man does not want to take the radical step of genetic and/or hormonal suppression of his dangerous traits, he will have to accept the risk of global disasters of his own making.

Closing words

At the start of our reconnaissance expedition through possible futures I said of our point of departure, the early years of the 21st century: 'We are doing better than ever before'. When I try to draw up the balance between the promises and dangers we encountered along the way, my feeling is that I am left with a positive net result. I have the impression that one hundred years from now somebody may again say something to the effect that life is, all things considered, better than ever before.

Of course, the pros and cons we discussed cannot really be cancelled against each other in an objective way. It is like comparing pears with apples, so readers may come to other conclusions. Some may find that I have often made the wrong estimate of which trend is likely and which is not. Others may generally agree with my estimates of probabilities and improbabilities, but still see a more bleak future than I do.

The question who is right is at this moment a wrong question. That will only be decided when the future becomes the present. I have stressed from the outset that *the* future does not exist. There is no predestination, only a range of possibilities. Some are much more likely to happen than others, but it is still truly undetermined which will finally become reality. All one can do when exploring the future, is to weigh up possibilities and point out trends. Even if I would have accomplished this perfectly (which I certainly do not pretend), it still could turn out in 2100 that my view of the future was utterly wrong. The simple reason is that something can be quite probable and yet not happen. For example, let us assume that in 2050 circumstances are such that possibility A has a chance of 90% to happen and possibility B a chance of only 10%. If it nevertheless is B that becomes reality, then history from the year 2050 onwards takes a different route than I had expected, although my calculation of probabilities was correct.

In short, one cannot do more than make an estimation of the probabilities of future developments. If one does that properly, one obtains a good prediction in the sense that there is a good chance of similarity with the way the future manifests itself when it becomes the present. However, until that moment it remains undecided in how far it was a 'correct' prediction.

The fact that there is no predetermined future means that it can be 'managed'. It can to a large degree be adjusted to man's needs by making the right choices at the right time. A change of course into the right direction is now taking place regarding population growth. There we are just in time. On the subject of global warming one has *not* done the right things at the right time. This does not mean that everlasting harm has been done, but it does make life more difficult for our grandchildren than was necessary. Maybe this book will contribute a little bit to the realization that we are shaping the future of humanity with our actions of today.

Reading suggestions

Good articles on about every topic and subtopic of this book can be found in Wikipedia. I list some book titles for readers who want to delve deeper.

Population growth, globalization and climate change

J. Randers, *2052: A global forecast for the next forty years.* Chelsea Green Publishing, 2012.

Globalization

T. Friedman, *The earth is flat: The globalized world in the twenty-first century.* Penguin Books, 2007.
J.A. Scholte, *Globalization.* St. Martin's Press, 2005.
M.B. Steger, *Globalization: A very short introduction.* Oxford University Press, 2003.

Climate change

A. Dessler and E. Parsons, *The science and politics of global climate change.* Cambridge University Press, 2010.
B. McKibben, *The global warming reader.* Penguin Books, 2012.

Politics and society

G. Friedman, *The next 100 years: A forecast for the 21st century.* Anchor Books, 2010.
K. Mahbubani, *The great convergence.* Public Affairs, 2013.

Science and technology

Quantum wonders:
J. Stolze and D. Suter, *Quantum computing.* Wiley, 2004.

Nanotechnology:
E.K. Drexler, *Nanosystems: Molecular machinery, manufacturing, and computation.* Wiley, 1992.
K. Hayles, *Nanoculture: Implications of the new technoscience.* The Cromwell Press, 2004.

Robotics and artificial intelligence:
R. Kurzweil, *The singularity is near: When humans transcend biology.* Viking, 2005.
N. Nilsson, *The quest for artificial intelligence: A history of ideas and achievements.* Cambridge University Press, 2010.
P.W. Singer, *Wired for war: The robotics revolution and conflict in the 21st century.* Penguin Books, 2009.

Genetic engineering:

G. Church and E. Regis, *Regenesis: How synthetic biology will reinvent nature and ourselves.* Basic Books, 2012.

A. Rutherford, *Creation: The origin of life / The future of life.* Viking, 2013.

Tinkering with ourselves

Brain research:

M. Nicolelis, *Beyond boundaries: The new neuroscience of connecting brains with machines.* St. Martin's Press, 2012.

S. Rose, *The future of the brain: The promise and perils of tomorrow's neuroscience.* Oxford University Press, 2005.

K. Warwick, *I, cyborg.* Century, 2002.

Better than nature:

A. Buchanan, *Better than human: The promise and perils of enhancing ourselves.* Oxford University Press, 2011.

J. Harris, *Enhancing evolution: The ethical case for making better people.* Princeton University Press, 2007.